Animal Housing and Human–Animal Relations

This book provides an in-depth investigation into the practices of animal housing systems with international contributions from across the humanities and social sciences. By attending to a range of different sites such as the zoo, the laboratory, the farm and the animal shelter, to name a few, the book explores material technologies from the perspective that these are integrated parts of a larger biopolitical infrastructure and questions how animal housing systems, and the *physical* infrastructures that surround central human–animal practices, come into being. The contributions in the book show in various ways how *physical* infrastructures of animal housing are always part of a much broader *sociocultural* and *political* infrastructure, where the material reality of housing systems combines with human and animal agents, with politics and with practices. As such, the book explores what kind of practices and relations develop around the physical structures of animal housing, and by whom, and for whom, they are developed. This innovative collection will be of great interest to students and scholars in animal studies, more than human studies, geography, anthropology and sociology.

Kristian Bjørkdahl is a Researcher at the Rokkan Centre for Social Studies, in Bergen, Norway.

Tone Druglitrø is a Post-Doctoral Research Fellow at TIK Centre for Technology, Innovation and Culture, University of Oslo, Norway.

Routledge Human–Animal Studies Series
Series edited by Henry Buller
Professor of Geography, University of Exeter, UK

The new *Routledge Human–Animal Studies Series* offers a much-needed forum for original, innovative and cutting edge research and analysis to explore human animal relations across the social sciences and humanities. Titles within the series are empirically and/or theoretically informed and explore a range of dynamic, captivating and highly relevant topics, drawing across the humanities and social sciences in an avowedly interdisciplinary perspective. This series will encourage new theoretical perspectives and highlight ground-breaking research that reflects the dynamism and vibrancy of current animal studies. The series is aimed at upper-level undergraduates, researchers and research students as well as academics and policy-makers across a wide range of social science and humanities disciplines.

Critical Animal Geographies
Politics, intersections and hierarchies in a multispecies world
Edited by Kathryn Gillespie and Rosemary-Claire Collard

Urban Animals
Crowding in zoocities
Tora Holmberg

Affect, Space and Animals
Edited by Jopi Nyman and Nora Schuurman

Animal Housing and Human–Animal Relations
Politics, Practices and Infrastructures
Edited by Kristian Bjørkdahl and Tone Druglitrø

Animal Housing and Human–Animal Relations

Politics, Practices and Infrastructures

Edited by
Kristian Bjørkdahl and Tone Druglitrø

Routledge
Taylor & Francis Group
LONDON AND NEW YORK

First published 2016
by Routledge

2 Park Square, Milton Park, Abingdon, Oxfordshire OX14 4RN
711 Third Avenue, New York, NY 10017

Routledge is an imprint of the Taylor & Francis Group, an informa business

First issued in paperback 2018

British Library Cataloguing-in-Publication Data
A catalogue record for this book is available from the British Library

Library of Congress Cataloging-in-Publication Data
Names: Bjørkdahl, Kristian, editor. | Druglitrø, Tone, editor.
Title: Animal housing and human-animal relations : politics, practices and infrastructures / Edited by Kristian Bjørkdahl and Tone Druglitrø.
Description: New York : Routledge, 2016. | Series: Routledge human-animal studies series | Includes bibliographical references.
Identifiers: LCCN 2015047618| ISBN 9781138854116 (hardback) | ISBN 9781315722337 (e-book)
Subjects: LCSH: Animal housing. | Human-animal relationships.
Classification: LCC SF91 .A55 2016 | DDC 636.08/31—dc23LC record available at http://lccn.loc.gov/2015047618

ISBN: 978-1-138-85411-6 (hbk)
ISBN: 978-1-138-54715-5 (pbk)

Typeset in Times New Roman
by FiSH Books Ltd, Enfield

Contents

Figures

Contributors

Beat Bächi is a Researcher at the Institute for the History of Medicine at the University of Bern (Switzerland). His current project deals with the agricultural, industrial and scientific production of psychotropic substances. He has mainly published in the fields of science studies, history of technology, and history of agriculture. Beat's contribution pertains to a wider project realized at the Archives of Rural History (Bern). Besides his interest in the history of cattle breeding he presently also works as a farmer.

Kristian Bjørkdahl is a Researcher at the Rokkan Centre for Social Studies, in Bergen, Norway. He is trained in rhetoric and cultural studies, and is currently working on a project on the communication of the swine flu pandemic in 2009. Much of his previous work has been in human–animal studies, in particular on the rhetoric of the animal movement, on changing representations of animal agriculture, and on animal science in popular culture.

Henry Buller is Professor of Geography at the University of Exeter. He has a longstanding interest in animal geographies with a particular focus on farm animals (including farmed fish) and farm animal welfare. Author of three Progress Reports on Animal Geography in the journal *Progress in Human Geography*, in addition to a number of articles and chapters on human–animal relations, he is a member of the UK Farm Animal Welfare Committee.

Tone Druglitrø holds a Ph.D. in Science and technology studies. She is a Post-Doctoral Research Fellow at TIK Centre for Technology, Innovation and Culture, University of Oslo. Druglitrø has published on the science and politics of laboratory animals and animal welfare, the professionalization of veterinary medicine and on writing radical laboratory animal histories. She is also the co-editor with Kristin Asdal and Steve Hinchliffe of the anthology *Humans, Animals and Biopolitics: The More Than Human Condition*, forthcoming on the Routledge series *Multispecies Encounters*.

Knut Fageraas is a Researcher at Norwegian Institute for Cultural Heritage Research (NIKU), Oslo. He is currently working on a Ph.D. thesis affiliated to the University of Oslo about the World Heritage Site of Vega in northern Norway with a focus on the productive process of creating a World Heritage site by institutional policies and how it affects the cultural production of local values, human practices and animal behavior. At NIKU he works primarily with issues on environmental management policies, processes of economic added value of heritage and expert versus non-expert interface. His research interests comprise cultural history, heritage studies, rural studies, nature–culture relationships and animal studies.

Irene Gustafson is Associate Professor at the University of California at Santa Cruz in the Film and Digital Media department. She teaches both production and critical studies courses, including courses in gender/sexuality studies, non-fiction theory and practice, and essayistic modes of production. Her film/video work has screened nationally and internationally and her writing has appeared in *Camera Obscura*, *The Journal of Visual Culture*, *The Moving Image* and others.

Dolly Jørgensen is an environmental historian with broad research interests, including animal reintroduction, medieval forestry and agriculture, medieval urban sanitation, the conversion of offshore oil structures into artificial reefs, and environmentalism in science fiction. She has co-edited two volumes: *New Natures: Joining Environmental History with Science and Technology Studies* (2013) and *Northscapes: History, Technology & the Making of Northern Environments* (2013). She was a practicing environmental engineer before earning a Ph.D. in history from the University of Virginia in 2008. She is Associate Professor of Environmental History at the Department of Ecology & Environmental Science, Umeå University, Sweden.

Robert G. W. Kirk is a Lecturer in Medical History and Humanities at the Centre for the History of Science, Technology and Medicine (CHSTM), University of Manchester. His work examines the role of nonhuman animals in human cultures, particularly within science and medicine, as well as the place of nonhuman animals in historical writing. Rob's contribution to this volume forms part of a wider study examining how, why, and to what consequence animal care and welfare became increasingly integrated within the biomedical sciences in the post Second World War period.

Lene Koch is Ph.D., DPhil., Professor of Medical Science and Technology Studies, University of Copenhagen, Department of Public Health. Koch's research has focused on the historical, social, ethical and regulatory aspects of recent biomedical technologies. Her work has included analyses of in-vitro fertilization, stem cells, genetic screening, prenatal diagnosis, genomics and the history of eugenics and compulsory sterilization and most recently the uses of animal models in modern biomedicine. Koch is a historian by training, but has expanded her methodological and theoretical interests to adopt broader social scientific approaches to capture the complex nature of the emergence of the new biomedical technologies. Her work has been conducted in the context of a range of international networks and includes both monographs and articles.

John Law is Emeritus Professor of Sociology at the Open University. He writes on: the performativity of methods; nature and culture; human–animal relations; and postcolonial forms of knowledge. He works empirically and ethnographically, and is currently writing about salmon and aquaculture.

Marianne Elisabeth Lien is Professor of Social Anthropology at the University of Oslo. She has published widely on food, marketing, consumption, food production, human–animal relations, salmon aquaculture and domestication. Her latest book is entitled *Becoming Salmon: Aquaculture and the Domestication of a Fish* (University of California Press 2015).

Christina Katharina May is a Research Associate at the Staatliches Museum Schwerin, Germany. Her studies mainly address the work of Marcel Duchamp and printmaking in the GDR. Despite her interest in conceptual and graphic art, her major field of expertise is the history of zoo architecture, spatial theory and landscapes of the 20th century, especially the relationship between natural sciences and aesthetics. She holds a Masters degree in Art History and Dramatic Theory from the Ruhr-University in Bochum.

Mette N. Svendsen is Associate Professor at The University of Copenhagen. In her research she explores how relationships, boundaries, and exchanges between human and animal are practiced in the development of medical knowledge and technology and how life is understood and administered in the interface between laboratory and clinic. Her work draws from anthropology, science and technology studies, and sociology to understand how biomedical knowledge production and its translation to the clinic are embroiled in existential questions of life's worth. She has published widely in anthropological and STS journals such as *Cultural Anthropology*, *Current Anthropology*, *Social Studies of Science* and *Body and Society*.

1 Animal Housing/Housing Animals

Nodes of Politics, Practices and Human–Animal Relations

Kristian Bjørkdahl and Tone Druglitrø

The various things humans do with and to other animals often depend crucially on the physical infrastructures with which we order and regulate movement. Just as there could be no Sea World without large-scale aquariums, stadium-scale seating, and plaque-infested pathways; and no lab animals without laboratories, animal houses, and cages; there could be no factory farming without battery cages, animal mass-transports, and conveyor-belt slaughterhouses. In the same way, "free-range animals" would be impossible without...a range.

Though it might be tempting to view the physical infrastructures that regulate the movement of animals as *mere matter* – as inert objects that *mean* nothing and *move* no one – we are better served by seeing them as complex nodes of politics, practices, and human–animal relations. This is the central point we want to convey with this book. Animal housing systems are always part of larger socio-cultural and political infrastructures; they are sites where the material reality of the housing system itself combines with the practices of human and nonhuman agents, with science, and with politics. If the physical things we use to house animals appear simple and straightforward, they are in fact precisely the opposite.

The battery cage is a case in point. Although in itself not much more than steel wire assembled in a particular way, the battery cage is also a symbol of agricultural development, a manifestation of the spirit of modernization that swept through the agricultural sector in the twentieth century. At the same time, the cage is much more than a symbol; it is a potent piece of material reality. As a material object, the cage structures how poultry farmers perform their job – one could even say it determines what it means to be a "poultry farmer," as it changes the farmer's job description from caretaker of sentient creatures to manager of machines. The cage also affects greatly the *number* of farmers who can stay in business. Because of the high capital investment needed to install a battery cage facility, this system creates an incentive to house a large number of animals, so that farmers can recoup their investment; this makes for a highly efficient production, but precisely for that reason, it also implicates the cage in the movement towards concentration of agricultural production – i.e. fewer, but larger, farms. As such, the battery cage becomes both a constituent component and an image of the central, not to say central*izing,* aspirations of agriculture in our time. Last but not least, the cage changes life for egg-laying hens – *dramatically*. The changes implied by the battery cage are thus quite immense. It is not that the cage causes all those changes, and certainly not that it inevitably does so. Rather, like most animal housing infrastructures, the changes implied by the cage have been constantly debated, negotiated, appropriated, and criticized. Whatever its proper place in the complex chains of cause and effect, the seemingly inert and uninteresting battery cage emerges – along with the other *things* that enable animal housing – as a complex node of politics and practices.

What makes the issue of animal housing an important matter of concern for the humanities and social sciences is made clearer with the example of the battery cage. It illustrates how "animal housing" is a central policy area that brings together issues of animal welfare, human health, cultural practices, historical developments, and different versions of "humaneness" – as well as different relationships between humans and animals. These are all pertinent areas of research for humanists and social scientists, and in recent decades, a broad scholarly movement has emerged which argues that we, by including both animals and things in our purview, can gain a richer methodological approach with which to study social phenomena. In rough alignment with this movement, the present book offers a compilation of empirically rich studies that in various ways seek to understand emerging relations and dynamics between the material and the discursive phenomena that make up animal housing structures and practices. Some of the questions that frame this book are: How do systems of animal housing, and the physical infrastructures that surround central human–animal practices more generally, come into being? What kinds of practices and relations are formed around these physical structures? What norms, ethics, politics, and values develop around animal housing practices? By and for whom are they developed? A feature common to most, if not all, of the texts in this book is thus a developed interest in the materiality of animal housing *structures*, combined with a tendency to see physical things as enmeshed in *practices* and *politics*. The contributions to this volume are in other words concerned with animal housing both as *noun* and as *verb*.

This starting point signals our connection to a research tradition that has been growing since the mid-1990s, which puts into focus "the complex nexus of spatial relations between people and animals" (Philo and Wolch 1998, 110). This tradition of human – or more correctly, *animal* – geography (see Wolch and Emel 1998; Philo and Wilbert 2000), did much to alert the growing field of human–animal studies to the centrality of physical structures in human–animal relations.[1] Indeed, as Henry Buller's recent inside "reports" from this field make clear, one will have a hard time sorting the inside of animal geography from its outside, as this has become a "field with increasing interdisciplinary connections" (Buller 2014a; see also Buller 2014b and Buller 2015). Alongside geography, anthropology is another social science discipline that has concerned itself with issues of space, place and housing, in the vast ethnographic literature on domestication. These studies have demonstrated how domestication has provided the particular narratives that we live by, and that it sustains, justifies, and makes legible the "particular historical trajectories and biosocial relations" that resonate in the

Euro-American context and tradition (Lien 2015, 5–6). As Cassidy (2007) points out, in *Domestication Revisited*, anthropology has been particularly concerned with the relationship between domestication practices and human mastery and control, and how it produces hierarchies and classificatory distinctions. She adds, however, an emerging insight in anthropology (particularly those invested in investigating human–animal *relations*) (e.g. Lien 2015): The unquestionable link between domestication and human mastery is far from obvious, and might not even be a particularly good description of domestication and housing practices. In other words, domestication is more than a question of unidirectional control. Yet, as Henry Buller's contribution to this book demonstrates, domestication often involves the infliction of violence upon animals, and thus can hardly be described as a mutual process between equal individuals.

If the nodes between physical structures, humans, and animals was ever an exclusive domain of animal geographers, that is no longer so. Likewise, if the issue of domestication was ever a sole concern of anthropologists (or archaeologists), they have been joined by other social science and humanities disciplines. Indeed, in this book, scholars with backgrounds from history of science, STS, rhetoric, anthropology, art history, cultural studies, and environmental history grapple with precisely this issue. They start, however, with an expanded version of the central insight of animal geography, namely that we should include animals – and things! – in social studies not just as metaphors, or as mirrors of humans, but as entities that "matter individually and collectively, materially and semiotically, metaphorically and politically, rationally and affectively" (Buller 2014a, 3). This rings in workable harmony with other notable developments in the humanities and social sciences, not least with perspectives within human–animal studies *and* STS, which highlight the roles that nonhuman animals, technologies, and things play in the project of ordering society (e.g. Clarke and Fujimura 1992; Bowker and Star 2000; Daston 2004; Barad 2003).

In the remainder of this introduction, we will sample what we believe are some helpful points of departure for the study of the nexus between physical structures, humans, and animals. These should not be taken to represent an overarching theory, but are best understood as overlapping perspectives and approaches. Also, our sketch does not amount to a framework that unifies all the various motivations, methods and ways of engaging with the issue of animal housing that are on display in this book, or that one finds in the social sciences and humanities more broadly. That would be an endeavor that exceeds the limits of this introduction. Thus, we have chosen to situate the issue of animal housing in developments that have emerged across various disciplines and can be linked to a variety of "turns" that are not easily separated, such as the "material turn," the "animal turn," and the "constructivist turn." Though they highlight slightly different concerns, these turns have in common a focus on how matter matters – how humans depend on, are shaped and challenged by, nonhumans; how human–animal relations are situated in constricted spaces; and how these relations take part in defining these spaces.

Drawing on these various perspectives and approaches we suggest, rather carefully, that the concrete sites and practices of animal housing are worth exploring in themselves, and we add only – what is admittedly quite significant – that a fruitful way to engage with this issue is with an assumption of humans, animals, *and* technologies as mutually constitutive phenomena, as entanglements that produce particular norms and politics of human–animal relations. Too often, discussions of "human–animal relations" are allowed to slip into the abstract, as if such relations had no placement in physical space, and no connection to the things one finds in such spaces. But as the essays in this book will show, such abstract relations, between what truly are two *immense* categories – Human and Animal – are often predicated on things, objects, physical structures, and material artifacts. In fact, many of the functions we ascribe, or rather impose, on animals would not even be possible – nor would the institutions and practices that crop up round those relations – if it were not for objects.

To the extent these objects have been at the center of attention, they have typically been cast as "instruments of torture," as what stands in the way of animal welfare. In what follows, we aspire to a more complex understanding of the various normativities that infuse the nodes of animal housing. As many of the subsequent chapters will show, the values that are produced in and part of animal housing practices are often in tension, such as that between use and protection, or between principles of efficiency and welfare – which means that the "good" cannot easily be sorted from the "bad." Instead of a simplistic critique that calls for "animal liberation," we suggest a richer understanding of how and why we house animals like we do, what the prerequisites and the implications of that are, as well as what it would take to imagine a different way.

The Agency of Things

In his book *Barbed Wire: An Ecology of Modernity,* Reviel Netz provides a creative narrative on the role of things in modern history. More specifically, through a history of barbed wire, he demonstrates the importance of things and technologies for demarcating space and controlling movement. Marx famously said that people, before they do anything else, must eat, drink, and clothe themselves, but in fact, says Netz, "even before people do those things, they must move, and they must occupy space." Consequently, "History takes place as flesh moves inside space; it is thus, among other things, about the biology of flesh – as well as about the topology of space" (Netz 2004, 228). As Netz shows, this regulation of movement takes place by the help of technology, and it orders not only human lives, but animal ones as well.

Netz is not alone in arguing for the importance of taking things into consideration. In fact, social science and humanities scholars have in recent decades come to a greater appreciation of the role of things in the construction of human and nonhuman lives. They have shown how things are not, as we have acquired the habit of assuming, passive, given, fixed, or neutral (Latour 2005). Lorraine Daston argues, for instance, that things – material, physical objects – have traditionally been thought of as speechless, thoughtless, without agency, and thus without import. But without things, she says, what would the world be like? Well, it would be "just a kind of porridgy oneness" (2004, 9). Without things, the world would lose distinction, and indeed, "Without things, we would stop talking. We would become as mute as things are alleged to be" (ibid.). Likewise, David Baird has advocated what he calls "thing knowledge," which refers to "a materialist epistemology … where the things we make bear our knowledge of the world, on a par with the words we speak" (Baird 2003, 39). It is not the case, argues Baird, that physical things are just props in a drama that is really about "words and equations" (Baird 2004, 4). "Instruments are not in the intellectual basement; they occupy the same floor as our greatest theoretical contributions to understanding the world" (Baird 2004, xvii).

Though they emanate from various backgrounds and highlight somewhat different concerns, these perspectives can, when taken together, stand as articulations of what has come to be known as "the material turn." In recent years, the material turn has been driven forth by fields such as cultural history, history of science, and science and technology studies (STS). The material turn is thus not a uniform movement, but if we could bundle up the different perspectives and traditions in the same basket and talk about a "new materialism," we would first have to note that this movement is characterized by a call to reconsider the ontological, epistemological, and political status of materiality; it is an agenda that champions the idea that matter is not passive or unitary, but active, forceful and plural (Lemke 2014, 2). If this turn has been particularly salient within animal studies, this is perhaps because animals have traditionally been seen precisely as a sort of "thing." In the tradition we inherited from Descartes, animals were famously thought of as *automata*. Humans were spirit, animals were matter; humans acted, animals were acted upon; humans were subjects, animals were – yes – *objects*. Consequently, in human–animal studies, the material turn has boosted efforts to dismantle the human–animal distinction that were already underway, and has allowed a newfound appreciation of other-than-humans and their potentiality as actors. One central aspect of these efforts has been the attempt to overcome distinctions between humans and animals based on speech. Thus, the material turn has been an effort to move away from the power of linguistics and has turned "every 'thing' – even materiality – into a matter of language or some other form of cultural representation" (Barad 2003, 801). The appreciation of materiality across the social sciences and humanities has been criticized, however, for an excessive concern with the matter of materiality at the expense of a concern for the role of language. "Discourse cannot be discounted!" say the critics. They suggest that we should attend not only to the role and agency of materiality, but to the semiotics that makes matter *matter* in particular ways, that enables matter to do stuff. In other – yes – words: When considering the thing, we also need to attend to the talk of the thing.

The Thing and the Talk of the Thing

The importance of acknowledging the co-construction of materials and language has been a central concern for scholars within science and technology studies

(STS) and particularly those working with Actor–network theory (ANT) (e.g. Latour 1987; 2005). The central tenet of ANT for instance – *radical symmetry* – emphasizes how all things (be it humans, animals, technologies, but also, importantly, speech and words) have the potential to be actors ("actants"), or to put it differently, have the potential to produce effects and outcomes. In this "material-semiotic approach," there is no such thing as *pure* materiality, since matter is always accompanied by scripts (Asdal and Ween 2014, 7). A crucial feature of material-semiotics, then, is its insistence that things and technologies are vital to the study of science, culture, and politics, but also, its insistence that these objects or technologies are always inscribed with meaning – with text and context. As such, technologies can be thought of as "speech apparatuses" (ibid.). This further entails that neither things nor their meanings are fixed or universal, even though we often think of them as such. At any point in time, what things are or can do is predicated on the networks they are part of.

What we should do, then, according to the material-semiotic approach of STS, is to pay closer attention to the work and practices of constructing and imagining arrangements (be it matter or meaning) for human–nonhuman spatial relations, and the politics and ideas that are part of these ordering practices. The notion of "infrastructural arrangements," introduced by Adele Clarke and Joan Fujimura in *The Right Tools for the Job* (1992), is useful to keep in mind in this context as it seeks to capture how infrastructures are crucially based on the theories, technologies, people and animals as well as ordering practices. Thus, the notion of infrastructural arrangements emphasizes the interrelationship between and coordination of heterogeneous elements in the facilitation of techno-scientific work. It involves an attentiveness to the work and craftsmanship of producing and adjusting organisms, technologies, instruments, resources, politics, theories, and methods when exploring how the ordering of human–animal relations and technologies produce identities, relations, skills, norms and ethics. It is exactly such contingencies and complexities of animal housing that this volume wants to tease out, and that the material-semiotic approach helps us to do. Allow us to elaborate.

The Politics of Animal Housing

If the assumption of radical symmetry offers one way of directing our attention towards the difference things make, and if it is also particularly apt for the study of animal housing, because both things and animals have been left out of the agentive realm, this has certainly not been the way that animal housing systems have always been understood. Half a century ago, Ruth Harrison published her ravaging critique of the emerging factory farms in the UK, *Animal Machines* (Harrison 1964), in which she displayed how much the lives of animals had changed in recent years. The transition from traditional ways to new, industrial infrastructures – not least those pertaining to the housing of animals – was a central part of her critique: "Farm animals are being taken off the fields and the old lichen covered barns are being replaced by gawky, industrial type buildings into which the animals are put, immobilized through density of stocking and often automatically fed and watered" (Harrison 1964, 1). Her story emphasized how animals were rendered as machines in factory infrastructures, how sentient creatures were thus made into (seemingly non-sentient) objects and commodities. Harrison's book was immensely influential, and is routinely credited with spurring the UK government to set down the Brambell Committee, the most famous result of which was the declaration of animals' "five freedoms" (Webster 2005, 12). The link between technological developments and the reduction of animal lives to *mere* matter is by now a well-known narrative. There is no denying the relationship between animal housing systems and violence; perhaps we even need to acknowledge this link to arrive at a responsible politics of animal housing. We do not, however, have to share the deterministic attributions that these criticisms often assign to the technologies and intentions of animal housing. In fact, it is our contention that we need to depart from the traditional view of technologies as reducing animal life to mere matter, and in particular, from the view that animal housing systems are nothing but instruments of domination. We thus need to rework our notions of technologies, to expand our view of what they can include, do, facilitate, produce, and represent.

In some cases, animal housing systems are most appropriately viewed as instruments of the human domination of animals, but in other cases, they must be seen as *nodes* of human–animal interaction, which implicate various *versions of care and management*. In still others, the quality or nature or function of the animal housing system is less easy to pin down. Hence, the politics of animal housing should not start from the *a priori* conclusion that all animal housing in itself is bad. This works as an argument for doing rich empirical studies that can reveal how animal housing structures and practices are results of ongoing appropriations and negotiations between heterogeneous actors. Further it works as an argument for getting at what STS scholar Hans Harbers (2005) has referred to as the "moralities, normativities and socialities of technologies." If we look for these aspects of technologies, we are also able to point to where crucial boundaries are drawn and where they are contested.

A prominent example of the complex of moralities, normativities and socialities that are performed and present in the farm animal industry is provided by Rhoda Wilkie (2010). Her account of the lives of farm animals differs markedly from the one Harrison gives. In *Livestock/Deadstock*, Wilkie illustrates the multifaceted and messy relationships between farm animals and the people dealing with them: The farm animals' status as commodities is an unstable one, and both instrumental and socio-affective relationships change according to the different settings of production, Wilkie writes (2010, 3). Vicky Singleton (2010a; 2010b) has furthered similar arguments in her study of British farmers and how they accommodate or fail to accommodate to the new British Cattle Tracing System (CTS), which was established in 1998. She shows how farming practices are primarily practices of responsibility and care that are embodied, relational, collective and responsive, and which cannot always be translated to the accounting systems established by legislative requirements, in an attempt to "purify" farm practices. There is a tension between local practice and systemic practice, Singleton argues. Thus, though they are often highly industrialized and, unavoidably, a part of global capitalism, farm practices depend on multiple modes of response. These modes are not necessarily calculable; rather, they have to do with coordinating nature and technologies to make things work and flourish in a "good" way, and this involves multiple forms of caring. A last example that we want to bring forth is John Law's (2010) take on care and killing in the farm animal industry. Based on ethnographic work in the aftermath of the foot and mouth outbreak in the UK in 2001, Law explores the enactment of multiple versions of care that were necessary to mobilize and enact, in order to respond *well* to, the plurality of ends at play: Care for the animal that is to be killed, care for the farmer that is losing his animals, care for the self, care for the veterinarian who performs the killing, *and* care for the national economy to which the cattle industry contributes.

Mobilized in such ways, the concept of care alerts us to the fact that the construction of technologies that manage life is not only a means of control *or* of liberation – even though this storyline has been and still is prevalent in both politics and fiction, not to mention sustained in public debate (Harbers 2005, 12). As the animal turn and other studies in the material and constructivist turn have shown, this is way too general, as it denies the mutuality between science, technology and culture, as well as the distributive relationship between human and nonhuman animal life. With their notions of care and of the relationship between local and systemic practices, Singleton and Law provide another take on the relationship between *structures and agency* when investigating animal husbandry practices. Their work challenges a view of technologies as uniform and stable, and emphasizes instead the need for empirical investigations that are attentive to contingency, complexity, and situatedness. Law's study, for instance, demonstrates that, when it comes to how the lives and deaths of animals are managed in animal housing practices, there are many values at play, at once, which all demand attention and concern; it is hardly ever a case of "one or the other." A last point that can be drawn from these examples on care and control is that they also challenge animal rights perspectives that are crucially concerned with the liberation of animals from human control. We need to be more sensitive to the plurality of concerns, values, normativities, and politics that are involved in housing animals.

By attending to animal housing as both a noun and a verb we can tease out such tensions and complexities without prejudging the *good* and the *bad*, and explore how norms and values emerge, are contested, and persist over time. Housing and domesticating animals are acts that crisscross normative boundaries of human–animal distinctions, and that reshape them, producing situated politics of bodies, agencies and relationships. The subsequent chapters engage precisely such sensitivities, and we hope that the compilation that results can demonstrate the need for more well-informed politics of animals housing.

Chapter Outline

This volume presents a collection of chapters that all, somehow, deal with the role played by the physical structures in which we house animals. A variety of sites and material objects are brought to the fore – cages, zoos, makeshift houses, labs, hospitals, shelters, underwater cylinders, boxes, slaughterhouses, free stalls, etc. – that mediate an equally large variety of human–animal relationships. The chapters display a similar variety when it comes to the scenes of housing, as they deal with homes, the public arena, commercial settings, medical and scientific sites, and conservation politics and practices. Further, the chapters demonstrate in their own specific ways that humans and nonhumans implicate each other, and how their joint histories also make versions of humans and of animals interdependent. In what follows we present a short overview of what you may expect from the various contributions. As you will soon realize, their cases, arguments, and stories exceed the focus of this introduction and further enriches the topic of animal housing.

The book opens with Marianne E. Lien and John Law's chapter on salmon aquaculture. It provides a particularly interesting case with which to think about what constitutes a house, more specifically, what work is needed to hold the "house" of fish together. As Lien and Law point out, even if the netting of the pens holds the salmon more or less in place, it performs rather poorly as a boundary in other respects: Sea-lice flow freely in and out; there are many microbes that cohabit the salmon domus, not to mention the humans that intervene. Lien and Law show how the salmon domus works as a "mediator" for knowing about and accommodating beings that are essentially different from humans – "we are nearly always apart, separated by the water surface," they write. Drawing upon new work on domestication in anthropology combined with material-semiotic perspectives in science and technology studies (STS), they are able to extend our conceptions and ideas of housing, what it can do, and how housing practices continues to mark turns in human and animal history of domestication.

In their respective chapters, Beat Bächi and Kristian Bjørkdahl show how animal housing innovations depend on the simultaneous reconfiguration of technology, animal bodies, and agricultural ideology. In other words, they tell stories that demonstrate how the development of farm technology often connects to transfiguration of the animal itself, and how those processes, in turn, tend to enliven debates about the nature – including the natural needs – of the animal. In his chapter, Bächi investigates the emergence of the "freestall barn" in Switzerland, and shows how this system was accompanied by controversies about the modification of the cow's body, notably the removal of its horns – which was needed to keep the animals from injuring one another. On the one hand, the freestall barn was simply a technological innovation, and an ingenious one to boot; now able to move around freely, the cows trampled their own manure down through specially designed gaps in the flooring, and in this way, relieved the farmer of drudging work. In the spirit of the times, this was seen as a move towards "rationalization" of animal agriculture. At the same time, the removal of the cows' horns – which was the basic premise of the freestall – had repercussions far beyond the stall itself, into Swiss culture and its ideas of what a cow ought to be. Was this dehorned animal still a cow – or was it some other sort of thing?

In Bjørkdahl's chapter, the backdrop is an established narrative that portrays the battery cage as a particularly cruel example of modern agro-industrial technologies – as an artifact that turned "sentient creatures" into "animal machines." While not denying that hens might suffer in cages, Bjørkdahl tells the story of the introduction of the battery cage to Norway, and finds that this process was more complex – even *troubled* – than one might think. The traditional notion of good poultry-keeping required the farmer to accommodate the natural needs of his or her animals. Partly for this reason, but also partly because of other factors pertaining to Norway's agricultural economy, the battery cage was well known – and discussed – for decades, before anyone began considering it a viable option. When that finally happened, it was a revolution – which not only affected the hen houses, but also the hens' bodies, as well as the poultry farmers' discourse. When that revolution was ultimately reversed – crowned with the EU's ban on the bare battery cage in 2012 – it had to do with the fact that the traditional notion – which was based on the idea that animals ought to express "natural behaviors" – was never really overturned.

In their chapter, Lene Koch and Mette N. Svendsen investigate the transfer of twelve capuchin monkeys from a psychiatric hospital laboratory in Copenhagen, Denmark to a small private zoo in the same city in 2004. Koch and Svendsen emphasize the complex politics involved in animal housing practices, and the contingent – often contradictory – ways in which animal lives are made to matter. For decades, the capuchin monkeys had been used in a variety of psychiatry related experiments. However, after massive pressure from animal protection actors and others, and due to a shift in the overall moral climate on animal welfare, a decision was made to "re-naturalize" the monkeys in the zoo. In their analysis, Koch and Svendsen unfold the different ways in which good housing arrangements are framed and how understandings of "naturalness" as well as human–animal relationships are co-constituted in the very same processes.

The complex politics and values enacted in animal housing practices are also very much the topic of Knut Fageraas' chapter on the traditional practice of housing eider ducks in the Vega Archipelago, on the west coast of Norway. Vega was in 2004 given the status of a World Heritage site. The eider ducks have been central to this area of Norway, and was also a crucial reason why this was given World Heritage status by UNESCO. Drawing upon resources from cultural studies and phenomenology, Fageraas shows how the World Heritage discourse takes part in purifying and reproducing particular notions and versions of animals and human–animal relationships, as well as particular versions of national culture and history.

Like Fageraas' chapter, Dolly Jørgensen presents a unique case in which the issue of animal housing is linked up to the politics and practices of conservation and cultural heritage. Jørgensen investigates historically the relocation and reintroduction of the muskox and the beaver into Norwegian and Swedish fauna. By focusing on a tool that seems quite mundane and unimportant – that of the *box* – she is able to attend to the issue of animal housing through practices of animal relocation. As Jørgensen points out, relocation containers are temporary, but in spite of the limited time of containment, these temporary houses are crucial points of transformation. The chapter argues that the box serves as a mediator that takes part in domesticating the animals as well as transforming them from beings that *come from somewhere else* to beings that *belong here*.

Aesthetics and design are central to building and constructing infrastructures, including animal housing facilities. In her chapter on the architecture of animal shelters in the San Francisco bay area, Irene Gustafson deals with the issue of aesthetics. She draws on feminist studies, animal studies and visual culture theory to frame her discussion. For Gustafson, aesthetics refers not only to a style of architecture, but also to the specific capacities for doing, feeling, and thinking that these spaces accommodate; she is concerned with what these spaces allow us to do, how they direct us in movement and action. Her chapter shows how animal shelters are rhetorically dense sites, in that they are designed to articulate how we imagine companion animals and their space and attachments to *home*. Her chapter reminds us of how objects orient us toward the world and how their "very thingness," as she writes, "stages an encounter and solicits a response."

The relationship between architectural design, encounters, and response is also the topic of Christina Katharina May's chapter. In a different geographical and historical context, May discusses the emergence of zoo design and architecture in Switzerland. She is particularly concerned with the work of the Swiss biologist Heini Hediger, who published one of the first scientific handbooks for keeping animals in zoological gardens in 1942. Interestingly, May shows how Hediger was reluctant to link the design of zoos to the discourse of aesthetics. He was concerned with building zoo infrastructures that facilitated individual animal behavior, served to educate the public on nature and animals, and enabled interaction between visitors and the animals. As May writes in her chapter: "By accepting contingency, each zoo visit would be unique, and the animals appeared to have the choice of being observed or not." Like Gustafson's chapter, May's story about Hediger and zoo design points to the power of mediation of culturally and historically specific values and identities through the building and staging of infrastructures around animals and how they relate to human sensibilities.

The issue of response, care, and human–animal relations has been a central concern in the rising body of work on laboratory studies in the history of science and STS. Both Robert Kirk and Tone Druglitrø concern themselves with the need for developing particular human–animal relations for the building of infrastructures to produce and house laboratory animals for use in biomedical science. They do so, however, in somewhat different ways.

Druglitrø traces the integration of monkeys in the production and testing of polio vaccines in Norway in the 1960s and 1970s and the care needed to "condition" the monkeys to work as part of the infrastructures of science. Druglitrø shows how experimental organisms in science are contingent upon the building of infrastructures that enable care practices that are specific to the animals housed. Further, drawing upon insights from feminist science and technology studies that have demonstrated how care practices are open-ended and depend on continuous tinkering, the chapter demonstrates how scientific work involves complex coordination of humans, animals, and various technologies and infrastructures.

While Druglitrø is concerned with understanding how particular forms of care becomes a central part of the work to construct infrastructures in the animal house, Robert G. W. Kirk investigates how distinct moral economies produced in the scientific animal house affect scientific objectivity in the laboratory. Mobilizing Lorraine Daston's (1995) concept of *moral economies*, he argues that ways of knowing in science operate through economies of affects and emotions that are necessary to "rational" scientific practices such as quantification and standardization. Focusing particularly on the construction of animal cages for the animal house, Kirk argues that these technologies shape the possibilities for multispecies concerns.

Mette N. Svendsen is also concerned with how relationships between humans and animals develop in the spatial arrangements of translational medicine. She does so by exploring ethnographically the qualitative difference between "housing" human infants and housing animal infants in neonatal research and clinical care. By mobilizing ideas of "biographical life" versus "biological life," Svendsen demonstrates how the care and management of piglets and infants is crucially linked to the purpose of a particular life, but also to the worthiness of different forms of life. The chapter exposes the different ways of enacting animal and human life in the medical context, but it also reveals the distinct differences between the clinic and the lab as medical sites, and, in addition, how these differences are articulated and enacted at scientific conferences. Svendsen argues that what becomes a qualified biographical life and a biological bare life is not socially predetermined or inherent in bodies, but an effect of socio-material practices and spatial arrangements. She points out that this perspective "direct[s] attention to the ways life is organized in confining certain humans and animals out in a field, inside a house, or away from cities or specific institutions."

As this brief summary of the chapters demonstrates, animal housing is a multispecies affair. It is not only about human control, but equally about human–animal modifications and modes of co-existence. As Henry Buller points out in the epilogue of this book, when we humans house animals it is a dual (or multiple) act, involving various versions of animalization and humanization; it is what Shukin (2009) has called a "zoo-ontological production." Further, the chapters demonstrate that "housing" is a complex, and far from self-evident, matter, and that it is performed with a range of means, to reach a variety of ends, which results in complex politics, practices and relations. To cite Buller (this volume), in conclusion, the nodes of politics, practices and relations that make up animal housing "derive from, yet also trouble, the relations we humans might have – or think we have – with sentient yet domesticated, nonhumans."

Note

1 The relationship between humans and things, and how this entanglement has played a crucial part in domestication processes have also been dealt with in archaeology, and particularly by archaeologist Ian Hodder (1990; 2012).

References

Asdal, K. and Ween, G. (2014) "Writing Nature," *Nordic Journal of STS. Special Issue: Writing Nature*, 2 (1): 5–10.

Baird, D. (2003) "Thing Knowledge: Outline of a Materialist Theory of Knowledge". In H. Radder (ed.) *The Philosophy of Scientific Experimentation*. Pittsburgh: University of Pittsburgh Press.

Baird, D. (2004) *Thing Knowledge: A Philosophy of Scientific Instruments*. Berkeley: University of California Press.

Barad, K. (2003) "Posthuman Performativity: Towards an Understanding of How Matter Comes to Matter," *Signs: Journal of Women in Culture and Society*, 28 (3): 801–833.

Bowker, G. and Star, S. L. (2000) *Sorting Things Out. Classifications and Its Consequences*. Cambridge, MA: MIT Press

Buller, H. (2014a) "Animal geographies I," *Progress in Human Geography*, 38 (2): 308–318.

Buller, H. (2014b) "Animal geographies II: Methods," *Progress in Human Geography*, 39 (3): 374–384.

Buller, H. (2015) "Animal geographies III: Ethics," *Progress in Human Geography*. Advance online publication: doi: 10.1177/0309132515580489 [Accessed September 25, 2015].

Clarke, A. and Fujimura, J. (1992). *The Right Tools for the Job? At Work in the Twentieth-Century Life Sciences*. Princeton, NJ: Princeton University Press.

Cassidy, R. (2007) "Introduction: Domestication Reconsidered," in R. Cassidy and M. Mullin (eds), *Where the Wild Things Are Now. Domestication Reconsidered*. Oxford: Berg.

Daston, L. (1995) "The Moral Economy of Science," *Osiris*, 10: 2–24.

Daston, L. (ed.) (2004) *Things that Talk: Object Lessons from Art and Science*. New York: Zone Books.

Harbers, H. (2005) *Inside the Politics of Normativity: Agency and Normativity in the Co-production of Technology and Society*. Amsterdam: Amsterdam University Press.

Harrison, R. (1964) *Animal Machines*. London: Vincent Stuart.

Hodder, I. (1990) *The Domestication of Europe: Structure and Contingency in Neolithic Societies*. Oxford: Blackwell.

Hodder, I. (2012) *Entangled: An Archaeology of the Relationships between Humans and Things*. Oxford: Wiley and Blackwell.

Latour, B. (2005) *Reassembling the Social: An Introduction to Actor-Network-Theory*. Oxford: Oxford University Press.

Latour, B. (1987) *Science in Action*. Cambridge, MA: Harvard University Press.

Law, J. (2010) "Care and Killing," in A. Mol, I. Moser and J. Pols (eds), *Care in Practice: On Tinkering in Homes, Clinic and the Farm*. Amsterdam: Transkript verlag, pp. 57–71.

Lemke, T. (2014) "New Materialisms: Foucault and the 'Government of Things'," *Theory, Culture & Society*, 32 (4): 3–25.

Lien, M. (2015) *Becoming Salmon; Aquaculture and the Domestication of a Fish*. Oakland: University of California Press.

Netz, R. (2004) *Barbed Wire: An Ecology of Modernity*. Middletown, CT: Wesleyan University Press.

Philo, C. and Wilbert, C. (2000) *Animal Spaces, Beastly Places: New Geographies of Human-Animal Relations*. London: Routledge.

Philo, C. and Wolch, J. (1998) "Through the Geographical Looking Glass: Space, Place and Society-Animal Relations," [Guest Editor's Introduction], *Society & Animals*, 6 (2): 103–118.

Shukin, N. (2009) *Animal Capital: Rendering Life in Biopolitical Times*. Minneapolis: University of Minnesota Press.

Singleton, V. (2010a) "Good Farming: Control or Care?," in A. Mol, I. Moser and J. Pols (eds), *Care in Practice: On Tinkering in Clinics, Homes and Farms*. Amsterdam: Transkript verlag, pp. 235–256.

Singleton, V. (2010b) "When Contexts Meet: Feminism and Accountability in UK Cattle Farming," *Science, Technology and Human Values*, 4 (37): 404–433.

Webster, J. (2005) *Animal Welfare: Limping towards Eden*. Oxford: Blackwell.

Wilkie, R. (2010) *Livestock/Deadstock: Working with Farm Animals from Birth to Slaughter*. Philadelphia: Temple University Press.

Wolch, J. and Emel, J. (1998) *Animal Geographies: Place, Politics, and Identity in the Nature-Culture Borderlands*. London: Verso.

2 The Salmon Domus as a Site of Mediation

Marianne E. Lien and John Law

Morten is squatting on the metal platform that connects the two rows of salmon cages, five on each side. At the far end there are offices and a warehouse stored to the ceiling with large plastic sacks of fish pellets. These are so heavy that you need a fork truck to move them to the hopper which feeds the automatic feeding system. Beneath his feet the water is shimmering blue. We are at Vidarøy, a salmon grow-out site in the middle of a fjord in West Norway. In the summer of 2012, Vidarøy is home to some 600,000 farmed Atlantic salmon, and during the day it accommodates 3–4 workers too, including Morten, who has been hired as a summer help. But what is he doing?

The answer is that he is holding a pair of scissors and a black garbage bag. He slides the scissors carefully along the full length of the bag, and repeats this until the bag has been turned into piles of long plastic strips, 10 cm wide. Then he makes each pile into a bundle, tying them together with a blue nylon rope coiled up in front of him. The rope is about 20 meters long and he's attaching four or five bundles of the piles to each meter of the rope. It is starting to look like a Tibetan prayer flag without the bright colors, but what is it for? "I am making shelter for the wrasse (*leppefisk*),"[1] he explains. "We are getting a new delivery [of these fish] tomorrow, and they need this, for cover. We will lower several ropes into each cage, with a metal weight at the bottom to hold them down and keep them steady."

Animal housing can involve permanent structures of the kind that are sometimes excavated by archaeologists; house-like arrangements that provide shelter against the rain and cold, as in the human–animal houses of the Viking era, or fence-like arrangements for livestock, such as the African kraal. But what is a house for fish? How does the salmon domus come into being? And how do material structures interweave with the relational practices of salmon farming?

Salmon Aquaculture

This chapter sees animal housing through the ethnographic lens of salmon aquaculture, and aquaculture through the lens of animal housing. Our approach draws inspiration from new work on domestication that seeks both to explore and to expand conventional notions of what this might be about.[2] But we do not want to propose a more refined or encompassing notion of domestication. Instead we take the latter, with all its indeterminacy and ontological baggage, as one possible analytical starting point from which to examine the heterogeneous practices involved in salmon aquaculture. The approach blends material semiotic versions of STS with what is often referred to as "lateral theory" in anthropology, but these labels are merely signposts for the reader. What matters to us here is what is going on with salmon as it unfolds ethnographically in the field. This means that our account is thoroughly situated. We draw extensively on joint fieldwork on and off salmon farms in West Norway,[3] where we have worked with people who work with the fish.[4]

West Norway is where the first experiments with salmon farming began nearly fifty years ago. Since the 1980s, salmon farming has expanded worldwide, and Canada, Chile, Australia (Tasmania), Scotland and the US all now supply salmon to the world market. But Norway remains the leading producer, shipping out more than a million tons of farmed salmon each year, mostly for export. In West Norway salmon aquaculture is a thriving industry, and has expanded until very recently. Indeed it has come to represent a threat to native salmon stock, as well as offering livelihoods and profit for local communities and global investors.

Salmon farming is at once a large-scale industrial enterprise and the most recent turn in the human history of domestication. While chicken, goats and pigs were domesticated thousands of years ago, and only gradually became subject to intensive mass production, Atlantic salmon are indeed "newcomers to the farm." But unlike chicken, goats and pigs, salmon farming was an industrial commercial enterprise more or less from the beginning. Small-scale freshwater aquaculture has been common in many parts of the world, and a significant contribution to diets, for example in China. But the enrollment of salmon in marine pens in the late 1960s marked the beginning of intensive marine aquaculture because it was scalable in the way that freshwater aquaculture was not. Hence, salmon aquaculture marks a turn, not only in the human history of animal domestication, but also of a simultaneous extension in the history of industrial food production from the terrestrial to the marine (Lien 2007a).

But the fact that salmon farming is embedded in transnational networks of commercial distribution and funded by capitalist investors does not define the day-to-day practices on a salmon grow-out site such as Vidarøy. Here, the salmon remain somewhat elusive, and mostly out of sight (Law and Lien 2013). Farm workers like Morten strive to keep them healthy and happy, through endless practices of counting, registering, watching and tinkering with a wide range of materials that help make the salmon visible or manageable. These practices (and the materials involved) cannot be separated from the structures that are holding the salmon in place, such as nets, pipes, and tanks, all of which constitute what we refer to as the salmon assemblage, or the salmon domus. Elsewhere, we have described this as a fragile assemblage of animate beings and inanimate things that loosely holds together and mutually constitutes the conditions of existence of its components parts (Lien 2015; see also Law and Lien 2014). In this chapter, we build on previous work while we turn our attention to the notion of salmon "housing" as material practice. Hence, focusing on the salmon domus, we are interested in what it consists of, how it gets made, and how far it extends.

Domestication at the Edge: Implications of the Water Surface[5]

The word domus derives from Latin and is conventionally linked to hearth and home, while domestication is often used to denote the process of adaptation in which something is converted for domestic use (tamed), or made to feel at home (naturalized).[6] This alludes to processes of transformation or mediation by evoking a distinction between something that is contained within house, household, or home, and that which is external, often also conceptualized as "wild." In this way, the domus enacts a boundary, a distinction between inside and outside. In animal studies, the domus is often translated to or equated with the animal house, the shed, the barn, the cage, the stable, the hive, or the corral. In the case of aquaculture, the most obvious parallel is the submerged nets and the floating structure together referred to as a cage (if square) or a pen (if circular). But even if the netting holds the salmon more or less in place, it performs rather poorly as a boundary in other respects. As we shall see, sea lice flow freely in and out, and pollock (which take up residence when they are young) find themselves trapped when they grow. Many other cross-species relations proliferate within the salmon domus too, including microbial entities in various forms of consumptive, collaborative or parasitic cohabitation. And then there are a few humans who try to make sense of it all and sometimes intervene.

The basic structure of the "salmon house," the circular pen, was first introduced in the late 1960s, after some early experiments with hexagonal structures. Since then, the size of the pens and cages has increased significantly, but the basic structure has remained more or less the same: a circular or square floating device around 30 meters across, holding netting which forms a cube, an oblong, or a cylinder. We will return to the nets shortly but first we want to touch on a key difference between aquaculture and conventional terrestrial husbandry.

One of the most important features of the salmon farming arrangement is the least visible: salmon spend their life in water. This, to state the obvious, is their medium. We often say that they are "below the surface," but that is of course a human perspective, and reflects the fact that our preferred medium happens to be air. Tim Ingold insists that "the medium affords movement and perception" (Ingold 2011, 149). We would add that it does far more. The medium makes it possible to breathe for mammals and fish: it conveys the oxygen without which life as we know it would be unimaginable. The air is not simply a medium. It also carries – is part composed of – oxygen. And there is oxygen in water too. But water is also crucial for movement. For salmon, the world is three-dimensional. The size of their cages is not 25×25 square meters, but 25×25×33 cubic meters. Vertical movements are as important as horizontal. Their's is a world in which pellets appear in abundance above and sink through the water. The temperature drops as depth increases too – as well as the light – and they can adjust their position to accommodate to shifting weather and seasons. In short their medium is unlike ours, they inhabit a different world. This means that on the occasions they come really close to us humans, (and for most salmon it happens only once or twice in a lifetime), they are nearly always out of water, flapping in a dip net or held firmly by a hand and uncomfortably gasping for oxygen. Or, perhaps, they are anesthetized.

We can see them, probably, more than they can see us. The fish we see from above are always only a tiny fraction of the fish that are actually down there. Looking down into the pen we see twenty, thirty, or maybe fifty, while there are likely to be 50,000 in the pen – a thousand times more. So this is the first thing to notice about salmon domestication: we prefer different media. We are nearly always *apart*, separated by the water surface. This means that the material arrangement that makes up the locality at sea is an interface that accommodates this basic separation. Perhaps we could say that the salmon domus is a device that both facilitates and mediates this spatial distinction between those that need water to breathe and those whose oxygen metabolism requires air.[7]

Many things follow: we need pressurized air to power the pumps that suck dead fish that fall to the bottom of the pen because we cannot get down there ourselves. We may need cameras to see what is going on. We need steel and platforms, railings and walkways keep us safely above water, but at the same time close enough to its surface to allow us to lean over the railings with a dipnet if we want to try to catch a fish. All this is to compensate for the fact that we, the humans, are species with lungs rather than gills, and therefore relatively clumsy participants in the three dimensional water-world of salmon. So we build our world around theirs. In a spot with ample depth and currents we create a structure for *us*: a house floating in the middle of the fjord, with moorings and a platform and cylindrical buoyancy devices underneath, but disconnected from most other things that make up a human community, and accessible only by boat. All of this so that the salmon can remain where *they* need to be, in *their* water-world, contained within the netting, and hence accessible to us even if they are mostly out of sight.

The salmon assemblage can therefore be seen as a complex material interface, a set of textures and affordances[8] that serves to negotiate insides and outsides, defining, situating and/or mediating across various boundaries. One such boundary is the water surface, signifying the two media that the laws of gravity hold horizontally apart. It doesn't need any help, it pre-exists practically everything there is on earth, including the species whose evolutionary development it shaped. The domus then, is situated right "at the edge," on the water surface, from where it allows further practices of separation, mediation and transformation.

Domestication as Confinement: Permeable Textures and Relational Skills

Let us turn to another surface, the netting that softly defines the edge of each pen or cage, outlining the limits of the movements possible for salmon – and the wrasse too. But as a firm boundary, it performs rather poorly, or perhaps we should say selectively. It allows the free flow of water – and this is needed. At the same time it lets in – and out – parasites such as sea lice, (and an occasional young pollock as well), while keeping schools of mackerel, cod and fully-grown pollock out. This partial enclosure allows many cross-species relations to proliferate both within and across the salmon domus. Such relations include not only salmon, sea lice, wrasse, and pollock, but also various microbes in a lively gathering of beings engaging webs of more-than-human sociality (Tsing 2013).

But the nets afford other relational possibilities too. Together they constitute a structure on which algae grow. The gradual growth of algae is called "biofouling," and one of the problems with biofouling on a salmon farm is that it slowly reduces the permeability of the netting: water flow falls, and hence the supply of oxygen too (Lien 2015). In this way, the netting, a seemingly passive tool of separation, is itself enrolled in a wider meshwork of opportunistic life. In other words, the netting is not only a tool that keeps lively entities apart, it is also an integrated part of that assemblage in that it allows new life forms to proliferate, and hence it is a generative dimension of the salmon domus. Anna Tsing describes similar generative practices as "worlding practices" and situates them as: "part of that design, intentional or unintentional, that gesture[s] towards the future, making worlds for the yet-to-come, as well as for the present" (Tsing 2013, 28–29; Haraway 1988).

So what do you do when what is supposed to be semi-permeable is gradually getting heavy with algae? More precisely: what do you do when that something is submerged in water, getting heavier every day, but remains necessary to hold salmon in place? There are various possibilities, but for square cages the answer is *tromling*. A brief explanation: If you look at the pen you will see that there are long metal cylinders running the full length on two of its sides. These cylinders – two meters high and twenty-five meters long – face one another across the pen. One has a whole lot of extra netting wound round it, enough for the whole cage, while the other is empty. The process of tromling involves winding the netting currently in the water up onto the empty cylinder, while letting out the dry netting on the other cylinder into the water. The net as a whole is twice as long as it needs to be, so from the point of view of the cage as a whole there is no difference. But at any given moment, one half of the net is dry while the other is in the water. The tromling swaps these around so that the algae growing on the net die in the air, hence the net becomes permeable again and water and oxygen can flow through the pen.

The algae on the netting remind us that the domus is a heterogeneous assemblage consisting of matter and materials both dead and alive, both human and nonhuman, and that the distinctions between these are not always clear-cut. The work that the netting performs unfolds along the temporal ebb and flow of biofouling (algae growth) and human labor (tromling). Each of these heterogeneous practices relies on and enrolls netting or the sun and wind to achieve their opposite futures: nets are either lively with algae, relatively impermeable and incompatible with salmon growth, or relatively "dead," permeable and conducive to salmon growth and well-being. In this way, the netting maintains its significance in the domus assemblage, while the domus itself is in a continuous state of flux, or becoming.

But to say that the domus consists of the netting, and that the netting needs to be cleaned now and then, is to overlook the corporeal effort and the skillful practices involved in performing the domus. Think again about tromling. We draw mostly on John's field notes, for the simple reason that tromling requires physical strength, and John's arms were stronger than Marianne's. (Or at least, that is what they both, and everyone else assumed). Tromling is hard work, and like many other tasks out on the salmon farm, it is gendered.

> You start by sliding the huge cylinder on its little rails to the edge of the cage. Next you tie three long ropes to the end of the net that you are going to wind in. Then you unhook the end of the net from the rail beneath the cylinder. To do this you need to pull on its edge, the rope along its edge, quite hard. Next you walk round to the sliding sides of the net. First you remove encumbrances, and in particular the leppefisk shelters, the strips of black plastic. Next it's time to unhook the sides of the net from the rail. Finally you need to winch up the pipe which collects the dead fish.
>
> Now everything's ready and you can start the tromling proper. You set the cylinder going and it starts to wind the wet net in. The theory is simple. It's like winding on film in a camera. But the practice is messier. First, the controls … are in the electric junction box way down the middle of each pen. You can't switch these controls on or off and be near the winch itself. In theory someone needs to stand there and work the controls. But this isn't practical because you also need to be: pulling the net in to help the cylinder get a grip on it; making sure that the net doesn't bunch up in the middle of the cylinder; and making sure that the ropes on the edge of the net don't get caught. You also need to make sure that the net doesn't slip off the deck and back into the water. So you need to be pulling in at least two different places, all the while looking to see what is happening as the net winds on to the cylinder.
>
> Then you finish winching the old "wet" net in and secure the cylinder in place by tying the raised net in three places to the deck. Now you need to reattach the net to the pen, and this takes some effort, especially near the corners. Then you re-tie the safety ropes beneath the deck level, which in places is pretty awkward since it is impossible to see what you are doing, or get both hands in the right position. All this done, you lower the wrasse protectors back into the pen. And, the final thing, you reposition the daufisk collector.[9]

Tromling is difficult. It is also difficult to write about. John notes that all sorts of things can go wrong. It takes two people, but a third pair of hands comes in handy. Together they can do about three cages a day, maybe four if they are lucky. If there is nothing else, something more urgent that requires their attention. During the last week of June 2012, and with John's help, we did the entire platform. And wrote about it too.

Anthropology can be notoriously ignorant when it comes to paying attention to the mundane heterogeneous practices that hold assemblages together, the seemingly insignificant instances of world-making that can sometimes create empires. If the Norwegian salmon industry emerged as successful in a globalized economy, it is due to many things that are beyond the scope of this chapter, but certainly includes endlessly repetitive and labor-intensive tasks of which tromling is an example. A focus on animal housing invites us to notice textures such as nets. But as John's account indicates, such textures are always relational. They take their performative shape in a series of unfolding practices. The marine pen was a great idea, an invention that made salmon aquaculture possible. But the nets are nothing if they are not enacted regularly, through textured relational practices, such as tromling.

Marcel Mauss is known for his work on reciprocity, but also for his attempts to fill in the "blank spots" left out or left open by his more famous uncle and mentor, Émile Durkheim (see Schlanger 2006, 10–13). His concern was with what we refer to here as heterogeneous material practices, or as he termed it then: *techniques*. Techniques were "traditional efficient acts" or (in a more elaborate definition from 1941/48): "An ensemble of movements or actions, in general and for the most part manual, which are organised and traditional, and which work together towards the achievement of a goal known to be physical or chemical or organic" (Mauss 1941/48).

"Traditional" denotes here something that can be taught, learned, or shared, and draws attention to the collective. "Efficient" is about making a difference or having an effect. What Mauss seems to have attempted here is to coin an analytical term to denote the more-than-human elements involved in human material practices. Unfortunately, the term *technique* never gained much traction in anthropological analyses, and was soon replaced by more human-centered terms, such as "skill." Cut off from the practices that render them lively, materials had become silent, analytically speaking, unable (as most things are) to act on their own. What Mauss was after is not unlike what we are trying to do here: to grasp the elusive, messy, almost magical moments when heterogeneous elements assemble in generative, performative practices. To understand salmon aquaculture, we need to pay attention to such moments, and to notice that they happen all the time, in mundane trivial tasks such as tromling. If the cage is a "salmon house," then tromling is about house-cleaning.

Towards the end of the day, Marianne came to help too. More pieces of rope were needed, and they had to be no more than two meters each. Marianne is 1.75 meters tall. Close enough. She got out a coil of rope and a pocket knife and started working:

> I like the stretching, and how my body measures on to lengths of rope, which in turn will hold heavy nets as they are doing the tromling. I am thinking of tromling as a kind of housekeeping. If too much algae grows on the net then there will be less light and less water coming through. Quite an effort has been spent on tromling today, doing cages 2, 4, 6, possibly even 8. The ones we prepared today were 11, 12 and 13, and they need doing by the end of the week. Actually, much of what goes on, which is not about feeding, is attending to the domus. Caring for it, and for the fish – indirectly.

The Salmon Domus as a Dynamic Relational Entity

So what is a salmon domus? How does it get made? We have noted that the salmon domus is located at the edge, mediating the air–water surface. We have seen how the netting is both a semi-permeable boundary for various microbiota, and an organic dynamic texture in and of itself. Maintaining this texture requires human efforts, *technique,* and reminds us that textures are also relational, performed in practices. But the salmon domus is nothing without its inhabitants. In the next section, we will turn to the salmon, but first we consider the wrasse. So let us return to that summer day in 2012, when Morten was busy cutting black plastic.

Morten creates a structure which is intended to provide shelter for the wrasse that cohabit with the salmon and the sea lice inside the cages. It wasn't always like this. When we began fieldwork in 2009, black plastic strips were not part of the salmon assemblage. Wrasse were less common as well, but buckets full of local and freshly caught wrasse were occasionally bought from local fishermen, and released into the salmon cages, as an experimental non-chemical way of controlling sea lice. The sea louse is a large family of parasites including *lepeophtheirus salmonis* that feeds off Atlantic salmon in salt water and proliferates around salmon farms. By 2010, the concentration of sea lice had risen, not only inside the cages of the Hardanger salmon farms, but also in the surrounding fjord, representing a threat to the young salmon smolt as they migrated from rivers down the fjord to the sea. This was partly due to the sharp increase in salmon production in the region, making close encounters between migrating smolt and farmed salmon (or wild salmon and sea lice) more likely. Hence, keeping sea lice numbers down was defined as a policy priority, not only for the health of the farmed salmon, but in order to protect wild salmon populations. At the time of writing it still is.

Wrasse like to eat sea lice. Or at least, that is how it appears. Their introduction to the salmon cages appears to hold the level of sea lice lower than would otherwise be the case. This means that they keep the farmed salmon healthier, and the surrounding waterways less affected by sea lice. This in turns means that they indirectly protect the migrating smolt as well. And with wrasse inside the cages, there is less need for other kinds of sea lice medication, which are costly, cumbersome, labor-consuming and in some versions a risk to the environment or worker health and safety.

In the beginning, wrasse, (sometimes sold by local boys from their small motorboats and offered to salmon farms for six kroner each), were simply released into the cages without any further ado. But gradually, as they became a more common part of the salmon assemblage, they became managerially visible too. Counting and reporting the number of fish in each pen is important on a Norwegian salmon farm, and is mandatory in relation to regulatory authorities as well. Previously, counts were only required for salmon.[10] In 2011, a new standard inventory form was added to the weekly report, listing the date, number, and specific species of the leppefisk delivered to each pen, the number of deaths, and their overall total. We could say that rather than being seen as ad hoc visitors to the cages, from that moment on they became officially registered salmon cohabitants. At the same time, the increasing demand for wrasse became a topic for marine biologists who started to worry about the sustainability of the local wrasse stocks. The possibility of raising wrasse was considered, and wrasse domestication trials were instigated. Veterinarians began to worry about wrasse welfare in the salmon cages too, and inexpensive innovations like the black plastic shelter were introduced and soon became a part of the cage inventory.

So why this elaboration about strips of black plastic? First of all, it tells us that *the salmon assemblage is a dynamic entity, continuously in the making.* No doubt this is true for many other versions of animal housing too, but it is particularly salient in the context of salmon aquaculture. After all, salmon are "newcomers" to the farm. Unlike terrestrial companions (whose histories with humans go back thousands of years and offer contemporary farmers a wealth of cultural scripts and material artefacts to copy or consider), contemporary salmon farmers have few such traditions to lean on. Like Morten, they make up the salmon assemblage as they go along, adding something here, changing something there, so that what we encounter in 2012 is not the same as what we learned to work with in 2009.

Other changes have taken place too. Some are celebrated and get registered as large and consequential: the invention of vaccines that rendered the use of antibiotics nearly superfluous; the invention of the dry feed pellet, which eliminated the problem of bacterial decay and which made fish feed storable as well as highly mobile. Other changes are more subtle and generally unnoticed by innovation research and financial investors, but they are also consequential. Indeed, during our four years of working in the field we registered a number of small, incremental, but significant changes of this kind. So, for instance, we watched a plastic container being retrofitted to help load dead salmon into the grinder with a fork-lift truck – an innovation that cut out quite a bit of heavy lifting. This was a work – indeed a triumph – of tinkering that involved: cutting out one side of the large plastic container in question; creating a hinge on which to hang the now separated side from a piece of piping; jury-rigging this with worm screw clips and wire; and then creating a mechanism made from a length of wood and ropes for holding the hinged side in place and releasing it at the appropriate moment. But this was just one example. We also saw the installation of a dimmer on a light switch inside a hatchery. This meant that the light could be switched on and off gradually. The problem was that the abrupt changes in the level of illumination – a simple on-off switch – startled the fish and stressed them. Then there were small but important jury-rigged changes to the cylinders used for tromling. In the way these were designed and installed, moving them meant that they regularly jammed. A set of ropes, blocks and pulleys put together by one of the workers shifted the vector of forces, so that the cylinders stopped jamming. And then, a fourth instance of the kind of changes created in tinkering was a series of small changes to the air pressure pipes for powering pumps that made connecting and disconnecting these easier.

What emerges from these examples is that maintaining the salmon domus is not only, or primarily, a matter of implementing a predefined set-up. Instead, the domus may be described as an assemblage that gets enacted through *fiddling, experimenting, and tinkering with the materials at hand.* This is our second point. Tinkering requires a level of human technological skill, a certain kind of craftsmanship, or what Mauss called *technique.* To some extent, such techniques are transferable across fields. If you get hired at a salmon farm it helps, for example, if you know about boats, and how to moor, work ropes and tie knots so that things stay in place in bad weather. But this is not enough, because the salmon farm is not primarily *about* boats and knots and moorings. It is about salmon. Tinkering at a salmon farm calls for other sensibilities too, such as the ability to pay attention to fish and pick up things going on that might be happening twenty meters below your feet. It is about imagining, predicting, and interpreting what salmon are up to, and what it takes to make them healthier, or happier, or easier to handle. We are saying, then, that the tinkering that renders the salmon assemblage dynamic and constantly changing, is a more-than-human practice.

And this is our third point: Tinkering at salmon farms is a *complex relational practice* that is not restricted to bilateral relations between humans and salmon. As our story about the black plastic strips indicates, it can include other species and entities too, such as wrasse and sea lice. An interface at the edge, dynamic, heterogeneous and multispecies – this is our version of the salmon domus.

Domestication as Rendering Visible: Mediating Surfaces and Mobile Pellets

Though what we have described is not all that different from what emerges in studies of terrestrial husbandry, the specificities are different. So, for instance, our fieldwork has taught us numerous ways in which humans take on active roles in knowing and accommodating salmon, but very little about such activities on the part of the salmon. Biologists tell us that over time, their bodies and behaviors change, and that they adapt to the salmon pen too. Experienced farmhands tell us about how they can tell one tank or pen from another, based on the behavior of that particular population of fish. But even so, the salmon remains elusive, like an image forever reflecting our own form in a cabinet of mirrors. How can we describe an "animal house" in which a key inhabitant remains muted, eyes fixed and without expression at least for us, swarming together in schools of many thousands, and almost always out of sight?

One way to navigate these murky waters is to pay attention to the practices that render salmon visible, the practical arrangements that make them known. As we have argued elsewhere (Lien 2007a; Lien 2007b; Law and Lien 2013), these are part and parcel of the salmon assemblage, indispensable to its structure and its stated purpose. It is by paying attention to these practices that we come to appreciate how a marine grow-out site differs from a terrestrial farm, or how aquaculture differs from agriculture.

So, for instance, Fredrik watches salmon. He engages in a particular way of watching, "*sjekke foringa*" ("to check the feeding"). This is usually done several times a day.

> I see him up above, he has climbed up to the narrow gantry above the platform, from where he gets a better view of the cages. In his left hand is a bucket full of pellets. In his right hand is a plastic scoop. Now and then he scoops up pellets from the bucket and throws them onto the surface. Then he remains silent, watching their response. But what does he look for?

Most of the time the surface reveals some activity, a kind of simmering like hot water almost on the boil. Occasionally, the surface is broken by a swift movement of silver in the air, visible only for an instant. If the sun comes at a certain angle we may get a sense of more silvery shadows moving below the surface. But a scoopful of pellets can sometimes create a visible effect. Even when the automatic feeders are distributing feed in a circular pattern (which they do most of the time) the salmon still respond. When they do, this is an indication that they are still hungry. It shows that they are probably doing well, and that it is safe to feed them a little more.

The person throwing pellets onto the water surface is not a passive observer. He is actively engaging in one of the key practices of salmon farming, a practice that requires skill and more than just a couple of weeks' training. "Checking the feeding" is a multisensorial, tacit and embodied way of knowing, one in which learning is inseparable from doing, and both are embedded in a context of practical engagement with the world.[11] The point is to spot a difference: are they behaving differently than they did yesterday? Are the cages different from one another? And if so, why would that be?[12] So "checking the feeding" produces locally and temporally situated knowledge in an environment cluttered with information. As a daily practice it is part of the textured relations that constitute the salmon domus as a dynamic entity; a co-constitutive practice in the enactment of what specific salmon *are*, at any given moment, at least in relation to human beings, as well as a reference point for future decisions. In this way, it also informs what salmon become.

At one level, salmon aquaculture is all about profit, and the job description for salmon is simple: putting on weight. Attending to the cost of production is key to a profitable enterprise, and in their warm and dry offices company directors will continuously ask themselves whether the way things are done is efficient for the purpose of producing salmon cost effectively. But out there in the wind and rain, on a platform in the middle of the fjord, the questions that the farm workers ask themselves are different, and have more in common with what humans have perhaps always asked themselves about the domesticated animals in their care: How are you doing today? Are you healthy? Is there anything I can do differently to make things better?

These may be questions common to husbandry but the *way they are asked* to salmon is new. Beyond touch, and with numbers in tens or even hundreds of thousands, salmon cannot be *felt* as individuals. What we have instead, are practices that render them visible "cage by cage," for instance in the form of collective feeding behavior. The pellets are indispensable to their growth, and hence to industrial cost and potential profit. But the pellets are also an invitation: it is the pellets that invite salmon to come to the surface. It is the salmon's appetite that replaces our sensory touch, our careful examination of the individual. The pellets help to make salmon visible.

If the water surface marks the boundary between our preferred media, water and air, then "checking the feeding" could be described as a practice that transcends or overcomes this boundary. But perhaps the surface is better described as interface. In this practice, and by means of a scoopful of pellets, the water surface is transformed into an interface capable of yielding specific forms of information. The water surface becomes then, not something that separates, but something that mediates or connects, and an active constituent of the salmon domus assemblage.

Extensions and Gaps: Domestication at a Distance

We have considered domestication from the ethnographic vantage point of a salmon grow-out site in West Norway, and paid particular attention to the textures and the relational practices that constitute the animal house, as it unfolds on or around the platform. Netting, tromling, and square cages are obvious components, but as we have seen, so too are wrasse, strips of black plastic, ropes and buckets of feed. The expression "animal housing" invites us to consider domestication as a kind of spatial confinement, as something done *in a place*. But as we have tried to indicate by attending to relational practices and interfaces, "salmon housing" does not do boundaries very well – or at least, only in ways that are selective. We have rehearsed this argument in our discussion of the nets. But the point we wish to make is broader, and concerns the extension of the salmon domus.

The wrasse are supposed to eat sea lice. Along with insecticide treatments, they help to hold down sea lice numbers inside the cages. But as we noted above, a main object of the current mitigation efforts is the migrating smolt, the wild salmon that reproduce in rivers on the West Coast of Norway. Farmed salmon can tolerate the presence of sea lice in moderate numbers, and it is the migrating smolt that are most at risk. This is one of the reasons why efforts – and coordinated efforts – to hold down the levels of sea lice are so important. This means that when Morten lowered the black plastic strips into the pen, he was not only accommodating the wrasse, and by default, the cohabiting farmed salmon. He was mitigating a health risk for the wild salmon too. Current regulatory measures rely on monitoring practices that consider entire watersheds. Hence, what Morten does at the salmon farm may have an effect upriver, many kilometers from the platform. But what then becomes of the domus? How far does it extend?

The warehouse is stored with bags of feed pellets. Every fortnight or so there is a new delivery. The fish meal and fish oil that make up the pellets came from the North Atlantic, and also from the South Pacific. An increased demand for fish pellets translates to an increased demand for Peruvian feed stock fish. As salmon farming expands, the pressure on Pacific feed stock populations is likely to increase too. The impact of such expansion is beyond the scope of this article. For present purposes, it is sufficient to note that the practices involved in maintaining the salmon domus have repercussions far beyond the immediate location of the grow-out site. From the fjords of West Norway to the shores of Peru, distant landscapes are enrolled in domestication practices, rendered vulnerable or altered as a result (see also Swanson 2015). In short, to think of salmon cages or nets as tools of confinement is not entirely wrong, but it is incomplete if it fails to see the many ways in which the salmon assemblage reaches far beyond the location of the platform.

Acknowledgments

We are grateful to our collaborators in the "Newcomers to the Farm" project, Gro Ween and Kristin Asdal, for helping us refine our ideas about domestication, and to Heather Swanson for alerting us to how domestication acts at a distance.

Notes

1 Leppefisk is the informal, generic Norwegian name for different types of wrasse that are used as cleaner fish at salmon farms in Norway. They include what is also known in Norway as *berggylt* (Labrus bergylta), *bergnebb* (Ctenolabrus rupestris) and *grønngylt* (symphodus melops).
2 See for instance Cassidy and Mullin (2007), Lien (2015).
3 The fieldwork took place in West Norway between 2009 and 2012, and was usually conducted jointly by Marianne Lien and John Law at fish farm sites on short one-week visits. Most of the fieldwork took place within one firm – we call this "Sjølaks AS" – and we are grateful to this anonymized firm for its kind agreement to let us locate our study on its sites, and for its additional generous practical support. We would like to thank all those who work for Sjølaks (they too are anonymized) for their warm welcome, their help, and their willingness to let us watch them at work. In many cases their kindness vastly exceeded any reasonable expectation or need. Additional interviews and observations took place with and in: other firms, appropriate government agencies, industry meetings and academic conferences.
4 For details, see for instance Lien and Law (2011) and Law and Lien (2013).
5 This section draws extensively on Lien (2015).
6 See for example Hodder (1990).
7 As we shall see, it mediates other human–salmon relations too, such as feeding, counting, and monitoring.
8 Affordances refer to "particular qualities of things through which an object lends itself to specific relational possibilities according to whether it is hard or soft, sharp or blunt, liquid or solid, pliable, malleable, or rigid" (Harvey and Knox 2013, 7; see also Ingold 2011).
9 Edited version of fieldnotes by John Law, 25–29 July 2012.
10 The weekly report lists fish deliveries, number of deaths (*daufisk)*, feed distributed, whether there have been escapes, and the incoming and outgoing numbers and weight of salmon for each pen/cage.
11 For a more detailed discussion of skilled vision in relation to aquaculture and feeding, see Lien (2007b). For a detailed ethnographic account of these practices, see Lien (2015).
12 Many things can interfere. If, for example, if the sun is high, or the water temperature is warmer than normal, then they tend to be deeper in the water, and it could appear that they are not hungry. If they have recently been treated for sea lice, counted, or

moved from one cage to another, this might account for a change in appetite. Checking the feeding involves human engagements with salmon across a translucent surface, as well as with memories of yesterday and the days before, considerations about the weather and other bits and pieces of information that might explain a difference.

References

Haraway, D.J. (1988) "'Situated Knowledges': The Science Question in Feminism and the Privilege of Partial Perspective," *Feminist Studies*, 13 (3): 579–599.

Harvey, P. and Knox, H. (2013) "Introduction," in P. Harvey, E.C. Casella, G. Evans, H. Knox, C. McLean, E.B. Silva, N. Thoburn, and K. Woodward (eds), *Objects and Materials*. New York: Routledge, pp. 1–19.

Hodder, I. (1990) *The Domestication of Europe: Structure and Contingency in Neolithic Societies*. Oxford: Blackwell.

Ingold, T. (2011) *Being Alive: Essays on Knowledge and Description*. London: Routledge.

Law, J. and Lien, M.E. (2013) "Slippery: Field Notes on Empirical Ontology," *Social Studies of Science*, 43(3): 363–378.

Law, J. and Lien, M.E. (2014) "Animal Architextures," in P. Harvey, E.C. Casella, G. Evans, H. Knox, C. McLean, E.B. Silva, N. Thoburn, and K. Woodward (eds), *Objects and Materials*. New York: Routledge, pp. 329–338.

Lien, M.E. (2007a) "Domestication 'Downunder': Atlantic Salmon Farming in Tasmania," in R. Cassidy and M. Mullin (eds), *Where the Wild Things are Now: Domestication Reconsidered.* Oxford: Berg, pp. 205–229.

Lien, M.E. (2007b) "Feeding Fish Efficiently: Mobilising Knowledge in Tasmanian Salmon Farming," *Social Anthropology*, 15(2): 169–185.

Lien, M.E. (2015) *Becoming Salmon: Aquaculture and the Domestication of a Fish*. Oakland: University of California Press.

Lien, M.E. and Law, J. (2011) "'Emergent Aliens': On Salmon, Nature and their Enactment," *Ethnos*, 76(1): 65–87.

Mauss, M. (1941/48) "Techniques and Technology," in N. Schlanger (ed), *Marcel Mauss: Techniques, Technology and Civilisation*. Oxford: Berghahn Books, pp. 147–153.

Schlanger, N. (2006) *Marcel Mauss: Techniques, Technology and Civilization*. Oxford: Berghahn Books.

Swanson, H.A. (2015). "Shadow Ecologies of Conservation: Co-Production of Salmon Landscapes in Hokkaido, Japan, and Southern Chile," *Geoforum*, 61: 101–110.

Tsing, A. (2013) "More-Than-Human Sociality: A Call for Critical Description," in K. Hastrup (ed.), *Anthropology and Nature*. London: Routledge, pp. 27–42.

3 What Is a Cow?

The Invention of the Freestall and How Cows Lost Their Horns

Beat Bächi

Introduction: The Cowshed as Analytical Space

Until the 1950s, most cattle in Switzerland were kept in cow barns "chained," which means individually fastened by a chain or a halter. Cattle kept indoors could lie on the ground or stand up to get fodder, drink water or get milked. Most of today's dairy cows live in a freestall. This new form of animal housing enables cows to move more or less as they like, at least within the borders of the cowshed and according to the rules of the farmer.

The case of the freestall seems to contradict received accounts of modern farming as an enterprise concerned only with dominating "nature." The freestall appears to cater for the needs of animals, and is thus a more "animal friendly" system. The materialization of "cattle friendly" views on animals as "instinct beings" in the freestall leads to the hypotheses that an ethological or "ethical" view of the cow does not merely pre-date or emerge in opposition to modern (dairy) cattle breeding, but, rather, was integral to it (see Woods 2007 and 2012). However, cattle living in a freestall usually no longer have horns. There was broad international consensus on the dehorning of cattle to avoid conflict and mutual harassment in the freestall. This was later contested by those who felt that dehorning cows was inconsistent with good welfare. So the question, "Is a hornless cow still a cow?," means not only, "Does it still remind us of a cow?," but also, "Can it still live out its natural behaviors?" Last but not least, it was not at all clear whether cows without horns were still cows in the framework of the herd book. The herd book is the official register or record of the individuals and pedigrees of a recognized breed of cattle. Accordingly, breeding organizations, as the keepers of the herd books for different breeds, had to answer the question whether cows without horns still had the right to be recorded in the herd book. Hence, this question also had political, economical, and legal dimensions.

Why did so many farmers since the 1950s invest a lot of money in building freestalls? The 1950s and 1960s are characterized by an intensified industrialization of agriculture. Thereby, the lack of workforce led to acute debates on the "rationalization" of agriculture. One of the main promises of the freestall was the reduction of work; not least by a substitution of human work by the work of animals. In this perspective, it was less the freestall and the cows' free movement from the cratch to the milking stanchion that was the central promise. First of all, the freestall itself disjoined spaces for feeding and milking and thereby opened up a space for movement that before was virtually the same place. Accordingly, it was much more the utopia of bringing cows to work in the cowshed and becoming agents of rationalization. It was in this context that the freestall started to interest Swiss farmers too. The Swiss case is chosen here, as thanks to the work of the archives of rural history, it is the only country where historical sources make it possible to study the emergence of the freestall in detail.

While Ruth Harrison's 1964 book, *Animal Machines*, was quite powerful for the animal rights movement (Sayer 2013), this movement would be the wrong angle from which to observe the actual nascence of the freestall barn. Or, as Bernhard Kathan puts it in his wonderful and stimulating (2009) book *Schöne neue Kuhstallwelt* ["Brave New Cowshed World"], certain elements of the promises of the freestall were quite general convictions, shared by broad ranges of society. Kathan discerns the introduction of the freestall (which is an example of a system whose conceptualization is due to the knowledge and experiences of ethology, mechanics, and informatics) as the main paradigm shift in cattle housing in the twentieth century. But, as Kathan mentions, whereas animal protection organizations for a long time bemoaned the "misery beard" on chained cattle, the success of the freestall is less a consequence of animal protection activities than of economic interests and technological possibilities. To quote Kathan: "The arguments of the animal rights activists say less about animals than about a society having unrestricted, free movement as a central illusion of salvation. Only a human being in motion seems to be healthy, able to succeed. Progress only seems to be possible if the freedom of movement is not restricted" (Kathan 2009, 234).

To analyze the "microphysics" of power (Michel Foucault 1977) materialized in the freestall, I want to cut back to one of the most important materials used earlier to manage cattle outdoors, namely barbed wire. Barbed wire was, so to speak, the direct opposite of the freestall. In the time of barbed wire, animals were seen to be "under control when they are deprived of their powers of activity" (Netz 2004, 15). Regarding the cultural history of barbed wire, Reviel Netz remarked that with the prevention of motion, force assumes a special kind of necessity: "Quite simply, being in a place is something you do with your body – nothing else – and therefore, to prevent your motion from one place to another, your body must be affected. The history of the prevention of motion is therefore a history of force upon bodies: a history of violence and pain" (Netz 2004, xi).

The freestall, like barbed wire, also incorporates power relations. Hence it is important to observe the development of the freestall in interaction with social and cultural change. But power relations materialized in the freestall are bound less to flesh and iron than to horns and concrete. The development of the freestall is a history of the relocation of physical power to a new art of self-governing. This new mode of self-governing was also one of the founding principles of supermarkets, where the consumers do the work previously done by the shopkeeper. Related developments can be depicted for example in restaurants and banks. More generally: Societies have become self-service societies (Bonhage and

Girschik 2005). In this environment, the freestall should enable cows in a certain way to become farmers in their own right. From this perspective, the freestall reveals both the impact of new materialities and topologies, and the cows' agency.

Where Does the Freestall Come From? From "Chained" Cattle to Self-Cleaning Devices

To reconstruct the history of the freestall, it is important to analyze who had the power to define what a cow is. To explain the power of breeding organizations in specifying what a cow is and how her appearance has to be, we have to go back to the end of the nineteenth century when new standards for "pure" breeds emerged. Concerning breeding cultures, the case of Switzerland is not so different from other countries. From the 1880s, there was a huge trend, supported by the federal state, to "purify" cattle. The same endeavor can be observed in other countries. As there is little research on the history of cattle breeding, I try to set the Swiss case in relation to Norway, where some work on this question has been done. Regarding the creation of Norwegian Red Cattle, Lars Christian Risan observed: "As a Norwegian state-controlled breeding program it was fully implemented during the 1890s, taking the form of 'purification' to correct earlier, unhealthy cross-breeding … Hence during the 1890s the large aesthetic variety of Norwegian cattle was formally ordered into race-clean regions" (Risan 2006, 144).[1] As in Norway, in Switzerland there was also a version of "pure-breed biology" and the doctrine of local adaptability. The discussions regarding pure-breed biology were thus tightly connected to the language of eugenics.

These standardization programs not only led to the creation of "pure breeds" but also to a certain centralization of cattle breeding. The breeding organizations became "obligatory passage points" for all kinds of knowledge and expertise in the realm of cattle breeding. Obligatory passage points can be thought of as the narrow end of a funnel or horn, which forces actors to converge on a certain question. It thereby mediates all interactions between actors in a network and defines the action program. The most important breeding organizations in Switzerland, and hence the obligatory passage points for cattle breeders, were the *Braunviehzuchtverband* (the breeding organization for brown cattle) and the *Fleckviehzuchtverband* (for speckled cattle), both established during the 1890s. As the keepers of the herd book and organizers of efficiency tests, they also were the most important centers of calculation in cattle breeding. For Norway, Risan has shown that by the beginning of the 1960s, the biggest customer for the new computers available then in Norway was the Norwegian Red Cattle Organization (Risan 2006, 147). Hence, a breeding organization had a lot of power in the process of defining what a cow is.

Switzerland was a latecomer regarding the material development of the freestall barn, primarily due to structural and environmental circumstances. Most farmers lived in a world divided into small sections, meaning also that livestock was quite restricted. The interest in the freestall arose from discussions about "open shed" versus "closed stable" for dairy cows. These discussions had originated in 1913 in the USA. The debates on open shed versus closed stable arrived quite late in Switzerland as the herds were so small that hardly anybody reflected on constructing completely new stables. What is central about these discussions regarding open shed versus closed stables for dairy cows is less on a material than on a theoretical level. The discussions triggered a new interest in the physiological mechanisms of animal motor behavior and featured a gateway for the conceptualization of cattle as homeostatic devices. There were discussions about the optimal temperature for cattle, their homeostatic mechanisms, and the effects on productivity. It was less a discourse of caution, more one of productivity. The less a cow had to "heat" or "cool down" herself, the more productive she seemed to be.

To allow not only the scientification of agriculture[2] but also material changes in the housing of cattle, the reflections on animals' homeostasis had to be coupled with the narrative of rationalization. In Switzerland, the strongest impact of Taylorization on the cowshed can be observed in the 1960s. At that time, the widely read Swiss agricultural review *Die Grüne* ("The Green One"), published by the *Schweizerischer Landwirtschaftlicher Verein* (Swiss Agricultural Association), dedicated a special edition to the freestall (see Rohrer 1960). In 1964, the annual conference of the Swiss Society of Agriculturalists/Farmers stood under the motto of "rationalization." Simply anything should become rationalized: The field, the household, and especially the farmyard. The need for rationalization (whatever this exactly meant) was underlined by recurrent hints of labor shortage. Through these debates the freestall was brought to Switzerland, at least discursively. But, as the Swiss cattle breeders' herds were simply too small, the freestall hardly offered rationalization potential. The literature usually calculated a stock of approximately 30 cows as the threshold to exploit this potential: such herds hardly existed in Switzerland at that time.

Slatted Floors and Cow Agency

Like in a true Tayloristic system, the farmers' expertise was still needed when it came to feeding. As the trained agronomist (a degree usually earned in Switzerland at the Swiss Federal School of Technology) Mathäus Rohrer explained in 1962 that "complete self-feeding" was not a possible solution as it would remove the influence of the "facility manager." Hence, it was much more the potential to make cows clean their own stables that boosted the attraction of the freestall.

To make cows clean their stables themselves seemed to be possible by another invention, that of "slatted floors" or "slat floor in pens." This device did not emerge in the cowshed, but originated in pig fattening. As pigs have a higher reproduction rate, pig breeding and reproduction had become industrialized before the cowshed was transformed into a factory. Furthermore, pigs were kept more densely than cattle. But as pigs demonstrated, the system of slatted floors could be used to bring animals to self-clean their stable floors. Pigs had become important workers in their own production. Why shouldn't cows pay in a similar way for their fodder?

Slatted floors are not passive elements in the production of animal machines. It seems that their *agency* was the main driving force for livestock dehorning or disbudding. Slatted flooring in pens is only "self" cleaning if there is a high density of animals. If the density is not high enough, animals evade (deliberately or not) tramping their own excrement through the slats. Or to put it in other words: slatted floors require livestock to be kept in tight clusters. On the other hand, high density puts the animal's health at risk. One of the prevention measures was the dehorning of cows to reduce accidents in the space of "free" movement. Or, as the architect Alex Stuber put it in the Swiss "Monthly Newspaper for Agriculture"[3] in 1964 under the heading of "animal health":

> The surface for an animal has to be reduced to the minimum, in order that the excrement does not remain on the floor ... Whereas the dehorning of cows is already recommended for the conventional freestall, especially in the case of brown cattle, this preventive measure is an inevitable prerequisite for the necessarily dense occupancy in a slatted floor stable.
>
> (Stuber 1964, 225)

Consequently, whereas in a conventional freestall it had at least been possible to leave the cows their horns by a more extensive way of keeping cattle, now this freedom evaporated in the gap between the grids. Now, the calculation of cows per square meter was no longer in the hands of the farmer (alone), but inscribed in the slatted floors and mathematical algorithms.

The slatted floors were constructed differently in every country. Stuber could show in his article the models of the German Democratic Republic, Austria, Denmark, United Kingdom, United States of America, the Federal German Republic, Switzerland, Norway, the Netherlands, Israel, and Sweden. In addition, slatted floors were produced from different materials, one of which was concrete. This led to the circumstance that one of the central propagators and translators of slatted floor stables in the German speaking countries became concrete – and the journal with the marvelous name *Beton*: "Concrete."

A central ideal tightly connected with the freestall, especially in its "denser" variation with slatted floors, was equality. It was clear in an agronomist's mind in the 1960s that the animals in a freestall had to be kept in small groups of 10 to 15 animals of the same age and sex. Cows in estrus and "nervous" animals should be separated (Stuber 1964, 225). Further advice was for especially "horny" bulls to be calmed down hormonally by the veterinarian (Koller 1968, 193). These hints for breeders demonstrate a new problem caused by the freestall: cows living in very hierarchical herds should be kept in equal peer groups, so cattle had to be even more standardized and normalized. Interestingly enough, the discourse was soaked with anthropomorphic metaphors. Inequalities and fights over positions in society should as much as possible become excluded from the cowshed through the production of homogenous cattle. For the breeder this implied, besides the hormonal "alignment" of cattle, the adjustment of the cows' reproductive cycles in order to (re)produce homogenous peer groups.

On the other hand, the freestall also led to a new kind of individualization and taking care of animals. In other words, the "self-cleaning" of the cowshed endangered the self-cleaning of the animals themselves. As Heinrich Wiesendanger pointed out in 1966, the manager of the cowshed had to ensure that the consistency or texture of the cows' excrement had to be as homogenous as possible and not too "thin." If the excrement's consistency wasn't within this normal range, the animals couldn't be kept clean (Wiesendanger 1966, 911). Furthermore, the keeping clean of the animals required individual boxes in the cowshed. As Wiesendanger formulated it: "A sufficient degree of the dairy cows' cleanliness can't be ensured and is impossible by an exclusive use of slatted floors in the whole cowshed (which means without boxes), because of the differences in dejection (liquid manure and the danger of diarrhea)" (Wiesendanger 1966, 912). Hence, in the emerging practices of keeping cattle in herds and viewing and analyzing cattle in terms of population genetics, this trend was accompanied by a trend towards individualizing cows. The practice is underlined by the implementation of individual boxes for the cow's own cleanliness and "leisure time." This introduction of a little bit of privacy by individual boxes should create a peaceful environment to optimize rumination and thereby enhance the production of milk.

Horns, Harassment, and Horny Cows: The Scientification of the Animals' "Individual" Needs

Science played a central role in the emergence of the freestall as we saw in the discussion regarding cows as homoestatic devices. Besides the scientification of agronomy, animal physiology and business economics, the emergence of ethology was of prime importance in the genesis of the freestall. A central door opener for ethology in cattle breeding was the introduction of artificial insemination.

To produce milk, cows have to give birth to calves. Artificial insemination together with the advent of the freestall created new problems for farmers. To detect the point at which a cow is on heat, e.g. when she is in estrus, used to be the sire's task. With the advent of artificial insemination it now became a very important and difficult task for the breeder to detect whether or not a cow is in estrus – and to order (deep-frozen) bull sperm. This task requires a lot of knowledge and time to observe and analyze the cow's behavior. The new problem of observing cattle in the 1950s and 1960s was accentuated by the less close relation between human and animal beings in the cowshed. Whereas farmers had much more physical contact with a chained cow, this proximity threatened to vanish in the freestall.

Also, another (hardly ever noticed but quite powerful) discourse entered the cowshed taking artificial insemination as a vehicle: The one about domestic animals' "psychosexuality." During World War II, women especially had started to think and write about animal sexual behavior. In the 1950s, men wanted to regain this field and started with a "scientific" approach towards animal sexual behavior, producing an enormous amount of literature. When in 1958 a special edition of the influential journal *Zuchthygiene* ("Breeding Hygiene") was published, Raphael Koller, the author of an introduction entitled "The Importance of Psychosexuality for the Reproduction of our Domestic Animals," could present more than five densely-printed pages of literature on this topic (Koller 1958, 193–216). This scientification of animal behavior is also mirrored in the use of new denominations. Whereas the realm of animal behavior before World War II was named *Tierpsychologie* (animal psychology), it became common during the 1950s to name this scientific discipline *Ethologie*.

The main starting points for these debates on domesticated animals' "psychosexuality" were reproduction or fertility research. Scientists had experimented since the 1920s with sexual hormones and gland transplantations to restore or improve the cows' and the sires' fertility. These debates became intermingled in the 1950s with artificial insemination and the "naturalness" or "artificiality" of reproduction in the cowshed. A point of discussion was the definition of animals as "instinct beings." During cohabitation, contractions of the uterus were observed in cows. But it remained unclear if cows felt these contractions as an orgasm. Nonetheless, the "individual needs" of cows had to be taken into account. Artificial insemination appeared to diminish the cattle's instincts in a dangerous way. It was said that this loss of instinct could lead to "anarchy" in the cowshed.[4] To prevent this "anarchy," there was research (especially at Bayer, Wuppertal) regarding the application of tranquilizers to cattle.

What was coined "anarchy" in the cowshed was tightly connected with the new phenomenon of "stress," until then mainly a manager's condition called stress disease.[5] In the cowshed, "stress" mainly meant mutual harassment, especially with horns. Stress and mutual harassment is not only dangerous for cows and men working in the cowshed, but also diminishes the production of milk. Furthermore, harassment with horns was an imminent threat to cows' fertility as it jeopardized their getting on heat. Hence, "stress" induced by horns was an economic problem of the highest importance in the freestall.

What is a Cow? The Breeders' Position on Livestock Dehorning and the Material Meaning of Horns in the 1950s and 1960s

There was much discussion about polling/dehorning cows in popular (agricultural) media especially during the 1950s and 1960s. What not has been studied yet is the point of view of breeders, and especially breeding organizations. As mentioned above, they were the central authorities in defining the breeding culture. These organizations were tightly connected with politics and were much more influential than one might think. Not least, as they were keeping the herd book, they had the power to define what a cow is – at least if she wanted to enter this record – and what she should look like.

The climax of the discussions in Switzerland on livestock dehorning for the freestall can be located between 1954 and 1967. The first letter of enquiry for dehorning cows arrived at the Swiss Brown Cattle Breeding Organization (Schweizerischer Braunviehzuchtverband, today Braunvieh Schweiz) in 1954. A breeder in Turgovia wanted to build a freestall, hinting at reasons of rationalization. He conceived the dehorning of his cows necessary to prevent mutual harassment in his new freestall. The breeding organization refused this petition with the argument that the dehorning would lead to an "unusual" appearance of "our Swiss brown cattle." Furthermore, dehorning was perceived as severe cruelty to animals. Another argument against dehorning was the importance of horns for age estimation and that it would hardly be possible to compare animals with and without horns at a cattle exhibition. Furthermore, until then, the herd book number was branded on the horn. To these more practical arguments, "psychological" points of view were added. These "psychological" aspects seemed to be especially important in a country like Switzerland with "traditional" breeders (Protokoll 418, 1954, 2).[6]

But there were also fewer "traditional" breeders in Switzerland. Hence it was the manager of the farm of the well-known Swiss footwear producer, Bally, in Schönenwerd, who participated in the debates concerning the dehorning of cows. In a letter published in "Die Grüne" in 1956, Heinrich Eckhardt, the manager of one of Bally's farms that owned about 40 cows, stated that to him and Bally it seemed rational to build a freestall. Besides all the advantages of the freestall, Eckhardt feared that the animals' "headdress" would bring a lot of anxiety that could degenerate into unbearable danger in the cowshed. To Eckhardt it was an open question why cows still had horns: "One has to ask to what end our cattle living in sheds still have horns, as their peaceful 'raison d'être' is to produce milk and meat. For this purpose, they don't need to be armed!" (Eckhardt 1956, 798). As the "natural" and "original" function of self-defense wouldn't exist anymore, and the animals just created damage and harm with their horns, he couldn't see a reason why one shouldn't do away with these "weapons." Hence, as most of the horns were shaped by the breeders in an artificial way, upwards and forwards, to Eckhardt horns had lost their "naturalness" and were products of the breeders vanity; even more dangerous than "natural" horns that would be more down and backwards. For him, horns were just a "useless externality." He advocated, instead of "externalities," more performance and health. Furthermore, he was convinced that farmers and breeders would soon get accustomed to the new appearance of their dehorned cattle. "One already has become used to completely different things." As he wrote of *wildgewordenes Kuhhorn* – what might be translated as "becoming wild cow horns" – the reader gets the impression that these horns were actors by themselves (Eckhardt 1956, 799).

To get some empirical material regarding a dehorned cows' behavior, Eckhardt had dehorned a cow "that could use their horns very well." As he observed, the cow completely changed her character. Now, deprived of her "arms of offense," she was totally peaceful towards her neighbors in the cow barn and the milker no more feared being wounded by the cow's horns. Eckhardt concluded his letter in favor of dehorning cows in the freestall by stating that – besides all the people out there, shaking their heads because of the dehornment – there certainly would also exist practically-minded farmers able to judge the cult around cow horns and their unpleasant side-effects in an objective way (Eckhardt 1956, 800).

The martial and military language of the Cold War structured discourses on dehornment. In Switzerland, there was also another (so to speak) "hot" war going on in the 1950s and 1960s, the so-called *Muni-Krieg*, the "Bulls' War" (Bächi 2014). This war was about artificial insemination. In Switzerland, breeding organizations (especially the brown cattle breeders) were strictly opposed to artificial insemination and saw themselves as advocates of "natural" reproduction. After the brown cattle breeders had lost the fight against artificial insemination that also became officialized at the end of the 1950s in Switzerland – until then artificial insemination was only allowed in special cases – the brown cattle breeders used even more drastic language to denounce dehorning. They spoke of a "cultural shame," contradicting the aesthetic sense of the breeder. Dehorning was said to be cruelty against animals and an "unprecedented" bestiality. Nevertheless, now there also emerged arguments in favor of dehorning in the midst of the breeding organization, too. For example, it was mentioned that eye specialists would advise dehorning (Protokoll 464, 1960, 6). Horns obviously became seen more and more in the framework of "risk."

Hence, as recently as 1959, a huge majority of the brown cattle breeders were still strictly against dehorning cows. Nevertheless, eminent exponents of breeding organizations recognized in that year that public discussions in newspapers initiated by animal rights activists had quite disappeared. One of the reasons for this was the small number of freestalls, as most of the farmers had quite a tiny herd. Furthermore, as purchasers of Swiss cattle in foreign countries still asked for cows with horns, the breeding organization repeated its refusal to record dehorned cows in the herd book, because cows without horns were seen as inconsistent with the "standard of the race." Also, 90 percent of the organization for the Simmental breed (the Fleckvieh breed) had advocated the proscription of dehorning. In 1959 there was only one single deputy amongst the brown cattle breeders who voted for liberalization in the question of livestock dehorning (Protokoll Fachmännerkonferenz 1959, 8–10).

Then, in 1964, an important change occurred. The Ministry for Agriculture reformulated the instructions for the right to have cattle recorded in the herd book: now it was the breeding value that was crucial. Furthermore, the breeding value was de-coupled from the animal's visual, external appearance and became an internal value, defined by performance (Protokoll 497 1964, 2). Hence, the horns as external elements of a cow became disconnected from the breeding aim that now became transmitted to the cow, expressed only by performance. But the breeding organizations didn't give in so easily. First of all, they appointed a legal expert with expertise regarding the question of whether a cow without horns was (still) a cow. Why did the breeding organization engage a legal expert to decide this question? Obviously, the right to register dehorned cows in the herd book as the central accounting instrument for cattle breeders had transformed the question of dehornment from a moral, ethical and aesthetic into a legal question. The legal expert, Prof. Liver, came to the conclusion – like the breeding experts – that a cow's head includes horns. But, as the chairmen of the breeding organization mentioned in the light of this piece of expertise, one could easily ask if horns "today still have the same material meaning" as they used to have (Fachmännerkonferenz 1967, 16). What obviously remained hegemonic was the characterization of horns as "weapons." In this perspective it was no mere coincidence that the discussions about horns could be brought together with serious debate in Switzerland at that time about "Mirage" aircraft – new fighter/bombers (Protokoll Kleine Fachmännerkonferenz, 1967, 17).

Conclusion: The Freestall as New Topology of Animal Husbandry

Several factors – the entry of ethology and new discourses on the "natural needs and desires" of cows and bulls, new material devices, cybernetics, and computer sciences – combined to create the freestall; and when they did, they exemplified a new "topology of animal husbandry," which was no longer predicated on the deprivation of movement and activity. Rather, in the freestall, a new – "lighter" and less visible – form of control originates, one that manipulates animals using ethological knowledge about their instincts, which integrates concerns for cows' welfare in modern cattle breeding. But in return, the cows had to give up something too – their horns.

Hence it doesn't make sense to tell the story of the freestall by a dichotomist opposition of animal rights activists and breeders, agricultural engineers, architects, business economists and so on. They shared a lot of general convictions and utopias. Freestalls and horns (and livestock dehorning) tell us a lot about social and cultural change. Polled livestock appears from this perspective as an expression of social and cultural transformation. First, they are signs of rationalization and the Taylorization of cow housing. Second, the freestall is a symbol for the utopia of freedom of movement and the association of continuous movement with healthy, productive, disarmed and peaceful beings. Third, slatted floors and the freestall barn are expressions of the phantasm of the self-service society. These processes were accompanied by a flexibilization of frontiers, norms, and power relations in many different social spheres. The frontiers, norms and power relations had become less hierarchical, dichotomized, and more permeable. Cows and farmers should work peacefully together, hand in claws, so to speak, to keep themselves and the slatted floors clean, and enhance productivity. Fourth, freestalls materialize the ideal of equality and balance. Now cows no longer were kept in hierarchical herds, but in peer groups. This also reinforced the normalization of reproductive cycles and the hormonal alignment. At the same time, when cows were reduced – especially by artificial insemination, quantitative genetics, and the re-definition of the breeding value as a matter of performance – to a mere part of the population, they were also individualized. To keep themselves clean and to enable peaceful rumination to enhance milk yield, they were given some privacy in their individual boxes: this should reduce the animal workers' stress and enhance bovine fertility. Fifth, dehorned cows on slatted floors are a product of the risk society – and the juxtaposition of disarmament in the cowshed and the arms race during the Cold War. They now posed a threat to humans and other, especially inferior, animals. On the whole, dehorned cattle in the free stall are a symbol for "soft disciplining," if one wants to bring Michel Foucault into the picture. Finally, the origins of cattle dehorning can be located in the space between the slatted floors. Or, as Tracy Chapman puts it in "Telling Stories": "There is fiction in the space between."

Archival Sources

Schweizerischer Braunviehzuchtverband, today Braunvieh Schweiz (Archiv für Agrargeschichte [Archives of rural history, Bern], AfA, Nr. 116), Zug, Schweizer Fleckviehzuchtverband, today Swissherdbook (Archiv für Agrargeschichte [Archives of rural history, Bern], AfA-Nr. 123), Zollikofen, Schweizerischer Verband für Künstliche Besamung, today Swissgenetics (Archiv für Agrargeschichte [Archives of rural history, Bern], AfA-Nr. 148), Zollikofen.

Acknowledgment

I am grateful to the collaborators in the "Agrarisch-industrielle Wissensgesellschaft" project at the Archives of Rural History, Peter Moser and Juri Auderset, and the Swiss National Science Foundation.

Notes

1 For the Swiss case, see Orland (2004), Moser (2010), and Weishaupt (1998); for the Netherlands, see Theunissen (2008) and (2012); for a comparison of Germany and the United States, see Matz (2012).
2 Regarding the emergence of an agrarian-industrial knowledge society, see Auderset, Bächi, and Moser (2012).
3 *Schweizerische Landwirtschaftliche Monatshefte*. This was the main interdiscursive journal in Switzerland in the field of agriculture where farmers, veterinarians, scientists, breeders, politicians, agronomists etc. published and discussed.
4 Ineichen Berchter, Bericht über die 8. Welser Fachtagung für künstliche Besamung, 3, in: Archivbestand Swissgenetics (AfA-Nr. 148), Dossier 001-08. This draft is more detailed than the text published in *Schweizerische Landwirtschaftliche Monatshefte*, 36 (1958): 451–452.
5 For the history of "stress" in general, see Kury (2012); regarding the invention of the stressed animal, see Kirk (2014).
6 For the following, see Schweizer Braunviehzuchtverband (AfA-Nr. 116), Bd. 3, 121-07: Vorstand, Protokolle 1950–1958; Protokoll der 418. Vorstandssitzung vom 25. Juni 1954; Bd. 759, 321-07: 6. Fachmännerkonferenz in Zollikofen 1959: Protokoll, Berichte, Thesen, Teilnehmerliste: 2; Protokoll der 438. Vorstandssitzung vom 16. Januar 1957; Protokoll der 464. Vorstandssitzung vom 10. August 1960; Protokoll Kleine Fachmännerkonferenz, 15. Februar 1967, Bahnhofbuffet Zürich-Enge.

References

Auderset, J., Bächi, B., and Moser, P. (2012) "Die agrarisch-industrielle Wissensgesellschaft im 19./20. Jahrhundert: Akteure, Diskurse, Praktiken," in B.

Brodbeck, M. Ineichen, and T. Schibli (eds), *Geschichte im virtuellen Archiv. Das Archiv für Agrargeschichte als Zentrum der Archivierung und Geschichtsschreibung zur ländlichen Gesellschaft*. Baden: hier+jetzt, pp. 21–38.

Bächi, B. (2014) "Der 'Muni-Krieg': Stiersperma und der Wandel der Zeit-Räume in der Viehzucht im 20. Jahrhundert," *traverse*, 2: 77–89.

Bonhage, B. and Girschik, K. (2005) "Die Selbstbedienungsgesellschaft. Banken und Einzelhandel in Zeiten rechnergestützter Dienstleistungen," *Akkumulation*, 21: 1–8.

Eckhardt, O. (Gutsbetrieb Bally) (1956) "Laufstall und enthorntes Vieh," *Die Grüne. Schweizerische Landwirtschaftliche Zeitschrift*, 28: 798–800.

Fitzgerald, D. (2003) *Every Farm a Factory: The Industrial Ideal in American Agriculture*. New Haven, CT: Yale University Press.

Foucault, M. (1977) *Discipline and Punish: The Birth of the Prison*. New York: Vintage Books.

Kathan, B. (2009) *Schöne neue Kuhstallwelt. Herrschaft, Kontrolle und Rinderhaltung* [Brave New Cowshed World]. Berlin: Martin Schmitz Verlag.

Kirk, R.G.W. (2014) "The Invention of the 'Stressed Animal' and the Development of a Science of Animal Welfare, 1947–86," in D. Cantor and E. Ramsden (eds), *Stress, Shock, and Adaptation in the Twentieth Century*. Rochester, NY: University of Rochester Press, pp. 241–263.

Koller, G. (1968) "Probleme der Kälberbeschaffung und der Aufstallungssysteme in der Rindermast," *Schweizerische Landwirtschaftliche Monatshefte*, 46: 187–195.

Koller, R. (1958) "Die Bedeutung der Psychosexualität für die Fortpflanzung unserer Haustiere," *Zuchthygiene, Fortpflanzungsstörungen und Besamung der Haustiere*, 2: 193–216.

Kury, P. (2012) *Der überforderte Mensch. Eine Wissensgeschichte vom Stress zum Burnout*. Frankfurt: Campus.

Matz, B.A. (2012) *Crafting Heredity: The Art and Science of Livestock Breeding in the United States and Germany, 1860–1914*. Dissertation, Yale University.

Moser, P. (2010) "Vom 'schönen' Stier zur 'eleganten' Kuh," in *Zeit und Stolz. Das Jubiläums-Magazin "50 Jahre swissgenetics."* Zollikofen, pp. 12–15.

Netz, R. (2004) *Barbed Wire: An Ecology of Modernity*. Middletown, CT: Wesleyan University Press.

Orland, B. (2004) "Turbo-Cows: Producing a Competitive Animal in the Nineteenth and Early Twentieth-Century," in S.R. Schrepfer and P. Scranton (eds), *Industrializing Organisms: Introducing Evolutionary History*. London: Routledge, pp. 167–189.

Rabinbach, A. (1990). *The Human Motor: Energy, Fatigue, and the Origins of Modernity*. Berkeley, CA: Basic Books.

Risan, L.C. (2006) "Whitehead's Philosophy of Unities Explored in a Case of Scoical Democratic Cattle Breeding," *Configurations*, 14: 127–156.

Rohrer, M. (1960) "Der Freilaufstall. Ziele und Möglichkeiten vom betriebswirtschaftlichen Standpunkt aus gesehen," *Die Grüne. Schweizerische Landwirtschaftliche Zeitschrift*, 49: 1455–1482.

Rohrer, M. (1962) "Ziele und Möglichkeiten des Laufstalles für Rindvieh unter schweizerischen Verhältnissen," *Schweizerische Landwirtschaftliche Monatshefte*, 40: 286–304.

Sayer, K. (2013) "'Animal Machines': The Public Response to Intensification in Britain, c.1960–c.1973," *Agricultural History*, Fall (2013): 473–501.

Stuber, A. (1964) "Laufställe mit Spaltenböden," *Schweizerische Landwirtschaftliche Monatshefte*, 42: 218–246.

Theunissen, B. (2008) "Heredity without Mendelism: Theory and Practice of Dairy Cattle Breeding in the Netherlands 1900–1950," in *Conference: A Cultural History of Heredity IV: Heredity in the Century of the Gene*. Berlin: Max-Planck-Institut für Wissenschaftsgeschichte, pp. 27–50.

Theunissen, B. (2012) "Breeding for Nobility or for Production? Cultures of Dairy Cattle Breeding in the Netherlands, 1945–1995," *ISIS*, 103(2): 278–309.

Weishaupt, M. (1998) "'Viehveredelung' und 'Rassenzucht'. Die Anfänge der appenzellischen Viehschauen im 19. Jahrhundert," in M. Fuchs and M. Weishaupt (eds), *Appenzeller Viehschauen*. St. Gallen: Typotron AG, pp. 11–48.

Wiesendanger, H. (1966) "Spaltenbodenlaufställe und Boxenlaufställe für die Aufstallung von Jung-, Mast- und Milchvieh," *Die Grüne. Schweizerische Landwirtschaftliche Zeitschrift*, 26: 909–935.

Woods, A. (2007) "The Farm as Clinic: Veterinary Expertise and the Transformation of Dairy Farming, 1930–1950," *Studies in History and Philosophy of Biological and Biomedical Sciences,* 38: 462–487.

Woods, A. (2012) "Rethinking the History of Modern Agriculture: Pig Production in Mid-20th Century Britain," *Twentieth Century British History*, 23: 165–191.

Further Reading

Agamben, G. (2003) *Homo sacer. Die Souveränität der Macht und das nackte Leben*. Frankfurt am Main: Suhrkamp.

Agamben, G. (2003) *Das Offene. Der Mensch und das Tier*. Frankfurt am Main: Suhrkamp.

Appadurai, A. (ed.) (1986) *The Social Life of Things: Commodities in Cultural Perspective*. Cambridge: Cambridge University Press.

Bächi, B. (2012) "Chemopolitik und Reproduktionstechnologien: Hormone, Vitamine und Tranquilizer in der Rindviehzucht (1920–1985)," *Blätter für Technikgeschichte*, 74): 93–113.

Giedion, S. (1984 [1948]) *Die Herrschaft der Mechanisierung. Ein Beitrag zur anonymen Geschichte*. Zürich: Ex Libris.

Green, N.K. (1962) "Rostböden für Viehställe," *Schweizerische Landwirtschaftliche Monatshefte*, 40: 72–75.

Haller, L. (2010) "Stress, Cortison und Homöostase. Künstliche Nebennierenrindenhormone und physiologisches Gleichgewicht, 1936–1950," *NTM (Zeitschrift für Geschichte der Wissenschaften, Technik und Medizin)*, 18(2): 169–195.

Haraway, D.J. (1991). *Simians, Cyborgs, and Women: The Reinvention of Nature*. London: Routledge.

Harrison, R. (1964). *Animal Machines*. London: Vincent Stuart.

Hartmann, U. (1978) "Kostengünstige und arbeitssparende Stallsysteme der Milchviehhaltung im Vergleich," *Schweizerische Landwirtschaftliche Monatshefte*, 56: 343–344.

Kolmanz, D. (2012) *Im Schweinestall der Moderne – Normierung, Normalisierung und Klassifikation moderner Nutzschweine*. Unpublizierte Diplomarbeit, Universität Bielefeld.

Latour, B. (1998) *Wir sind nie modern gewesen. Versuch einer symmetrischen Anthropologie*. Frankfurt am Main: Fischer Taschenbuch Verlag.

Latour, B. (2010) *Das Parlament der Dinge. Für eine politische Ökologie*. Frankfurt am Main: Suhrkamp.

Polge, C. (2007) "The work of the Animal Research Station, Cambridge," *Studies in History and Philosophy of Biological and Biomedical Sciences*, 38: 511–520.
Risan, L.C. (2009) "La pratique de la social-démocratie à travers l'élevage de bovins," *Ethnologie Francaise*, 2: 341–351.
Rohrer, M. (1979) "Betriebswirtschaftliche Betrachtungen zur Wahl von Stallsystemen," *Schweizerische Landwirtschaftliche Monatshefte*, 48: 275–292.
Roscher, M. (2011) "Where is the Animal in This Text? Chancen und Grenzen einer Tiergeschichtsschreibung," in Chimaira – Arbeitskreis für Human-Animal Studies (eds), *Human-Animal Studies. Über die gesellschaftliche Natur von Mensch-Tier-Verhältnissen*. Bielefeld: transcript, pp. 121–150.
Rose, N. (2001) "The Politics of Life Itself," *Theory, Culture and Society*, 18(6): 1–30.
Schreiber, C. (2007) *Natürlich künstliche Befruchtung? Eine Geschichte der In-vitro-Fertilisation von 1878 bis 1950*. Göttingen: Vandenhoeck & Ruprecht.
Schrepfer, S.R. and Scranton, P. (eds) (2004) *Industrializing Organisms: Introducing Evolutionary History*. London: Routledge.
W. B. [Full name unknown] (1952) "Die Offenstallhaltung in der Schweiz," *Schweizerische Landwirtschaftliche Monatshefte*, 30: 80–81.
Wadiwel, D.J. (2002) "Cows and Sovereignty: Biopower and Animal Life," *borderlands*, 1(2).
Werner, F. (2011) *Die Kuh. Leben, Werk und Wirkung*. München: Goldmann Verlag.
Wilmot, S. (2007a) "From 'Public Service' to Artificial Insemination: Animal Breeding Science and Reproductive Research in Early Twentieth-Century Britain," *Studies in History and Philosophy of Biological and Biomedical Sciences*, 38: 411–441.
Wilmot, S. (2007b) "Introduction: Between the Farm and the Clinic. Agriculture and Reproductive Technology in the Twentieth Century," *Studies in History and Philosophy of Biological and Biomedical Sciences*, 38: 303–315.

4 When the Battery Cage Came to Norway

The Historical Path of an Agro-Industrial Artifact

Kristian Bjørkdahl

For a long time, the Norwegian Egg Centrals – a cooperative of egg producers, and the main actor in this market – packaged and sold its eggs in a controversial carton. A rustic drawing rendered an old-fashioned, red-painted farmhouse placed on a grassy yard, on which three hens were strutting about under the bright sun. Aptly – or so it seemed – the contents of the carton carried the moniker "Sun Eggs." The controversy first arose when the vocal animal activist Kåre Knutsen, in his book *Dyrenes rettigheter* [The Rights of Animals] (1974), launched a scathing critique of industrialized egg production, laying bare the somewhat disingenuous contrast between the romantic representation on the carton's exterior and the actual origin of the eggs inside. In reality, Norwegian egg-laying hens were now predominantly kept in battery cages – long columns of tight metal wire cages stacked several rows high, operated not so much by human beings as by machinery. This was an environment, noted Knutsen, that afforded the animals little space in which to move about, and which thus prohibited natural behaviors and, therefore, animal welfare. On the egg carton, in contrast, "[e]verything smells of sunshine and health," and the message conveyed was one of hens "walking leisurely about, nibbling on grain and straw."

> Do the Egg Centrals not agree that the ads on these cartons are rotten? The hen strutting about at liberty can today be found only on these cartons and in children's ABCs. In this most modern of worlds, hens never get to see the sun, but are kept indoors their entire lives. Inside – where there is often very little light – she can exert her full efforts on laying Sun Eggs.
>
> (Knutsen 1974, 25)[1]

Knutsen's indictment could do little to remove the carton from the shelves, and nothing at all to rattle the battery cage system of egg production, which at the time of his writing had become the dominant one in Norway. Sun Eggs remained in sale, their disingenuous wrapping intact. Outcries against the incarceration of egg laying hens multiplied, however, fronted not least by the so-called Campaign against Caged Hens. Throughout the 1970s, this campaign assembled petitions, organized stands on Oslo's main street, and issued a steady flow of op-eds to the newspapers, in an attempt to inform the public of how egg-laying hens actually lived, as well as of how *unnatural* those lives were. In a typical sample of their complaint, a frequent contributor of texts on behalf of the Campaign wrote:

> It stands on a slanting wire floor, and lives its entire life on a space not much bigger than the size of a sheet of paper. It never sees sun or grass, but is exposed exclusively to artificial light … It develops deformed claws, and when a cage hen is taken out and placed on the ground, it cannot walk! Cannibalism is common, and weak animals are often picked on until they die … A caged hen cannot pick in the ground for feed, stretch its wings, ruffle its feathers, sandbathe, or develop social standing in normal ways. It cannot hide in a soft place when laying its eggs, or find a place to perch and rest.
>
> (Gleditsch 1976, 9)

As with Knutsen, the Campaign's work seemed to be of no effect. In 1981, the daily newspaper *Aftenposten* informed its readers that the issue of caged hens had been a "never-ending event in Aftenposten's letters column," and added that: "At this point, we cannot continue this repetitious debate. The space in the columns is too scarce, and the contributions too many" (N.N. 1981, 23).

In 1996, the attention of the critics turned once more toward the infamous carton, as the Norwegian Society for the Protection of Animals filed a lawsuit against the Egg Centrals, claiming that the image on the cartons should be considered "misleading advertising." *Aftenposten*, now willing to report on the issue once more, wrote that,

> The Society for the Protection of Animals has sued the Egg Centrals, demanding that the image of the happy hens must be removed, or replaced by an illustration which gives a more correct impression of how the eggs are produced, that is, of caged hens living in miserable conditions behind bars in Norwegian hen facilities.
>
> (Bøhm-Pedersen 1996, 6)

In the end, the court ruled that the advertisement was in fact misleading, and demanded that the image be replaced. The Egg Centrals' brand, Prior, reluctantly complied. They kept the rustic drawing in place, however, and only swapped the free-ranging, grain-nibbling hens with … an egg.

The Rise and Fall of a Cruelly Efficient Technology

In recent decades, the issue of caged hens has become an increasingly hot topic in Norway – as it has also elsewhere – not least because of the EU's widely publicized ban on the bare cages in 2012 (EU 1999; see also Appleby 2003). Largely thanks to this ban, the popular story about the battery cage is now often rendered as a narrative of "rise and fall." The battery cage, so the story goes, was a prime example of the cruelly efficient technologies introduced by the relentless forward-marching process called "industrialization of agriculture," a process that makes animals into machines and farms into factories (see Harrison 1964; Singer 1975; Davis 2009). Over time, the battery cage became a sort of reverse totem for critics of modern animal housing systems, a symbol of the animal abuse entailed by modern agricultural technology. The unique feature of this particular technology, however, was that it, apparently, was sent to its grave by popular opposition. When the EU ban went into effect in 2012, reactions were grand and self-congratulatory. Many newspapers mistook the event for a ban on cages altogether. One Norwegian newspaper declared, for instance: "All hens are liberated!" (N.N. 2011). Meanwhile, the Norwegian meat and dairy cooperative, Nortura, trumpeted: "Caged hens – now soon a thing of the past!" They added – arguably in a move of profound euphemistic impertinence – that Norway's three million hens would now be relocating to "environmental interiors" (Nortura 2011).

Although not necessarily wrong, this narrative is dangerously simplistic. It does not account for the complex web of incentives into which this agricultural artifact has been – and is still – entangled. Rather, it reduces the question of the battery cage to a simple struggle between good and evil – the *bad* cage that bereaved hens of the chance to fulfill their needs against the *good* forces (animal activists, consumers, the EU) that protested and ultimately put an end to this particular form of animal abuse. But this Manichean portrayal makes us none the wiser. In fact, we might ultimately be doing ourselves a disservice by using this sort of victory rhetoric, since the battery cage is in fact not even *banned*.[2]

Instead, we should be asking questions that probe deeper into the very existence of this technology. If we do so, we will discover that, at least in the case of Norway, the battery cage had to get past a widespread notion of good poultry keeping that was completely contrary to what the battery cage technology entailed. The traditional idea was that hens needed easy and frequent access to the outdoors, where they would have plenty of fresh air and sunlight, as well as ample interaction with conspecifics, not to mention the chance to sandbathe, churn and kick, and search for food. In short, it was believed that a hen needed to live out its *natural behaviors* in order to thrive, and that it needed to thrive in order to lay eggs (see Berger 1997). One central part of this notion was the emphasis on *movement* – an emphasis that obviously could not be satisfied by the battery cage, the most characteristic feature of which is its ability to *restrict* movement, and thus, behaviors.

But if that is so, why and how was the battery cage introduced to Norway in the first place? Given that these were so deeply rooted notions of what constituted a good life for an egg-laying hen, how could this apparently opposite technology – one that denied hens all the requirements of a natural life – work itself into Norway's agricultural system? In other words, if there was already a consensus that hens ought not to be incarcerated, why were they after all so thoroughly incarcerated?

In an attempt to counter simplistic narratives about the battery cage, I want in this chapter to show that the introduction of the cage to Norway was a slow, in many ways troubled, and anything but predictable, process. There was nothing automatic or inevitable about the spread of the battery cage, nor was there anything like a "rhetorical showdown" from which the battery cage emerged as victor. Instead, the introduction of the cage on Norway's egg farms only gradually became a new premise for the debate – a premise that at the same time began to be challenged from the outside. When that happened, the community of egg producers put aside their differences, consolidated, and went into defense mode. They had already gone too far towards marginalizing the traditional notion of good poultry keeping, however, to enable any meaningful exchange with the critics – who still started from the assumption that hens needed to live *natural lives*. The history of the battery cage thus illustrates, I argue, the increasing distance – geographical, sociological, rhetorical – between producers and consumers of agricultural products. At least in the context of this chapter, the battery cage is an example not so much the evils of industrialized agriculture, but of a particular form of *agricultural alienation* that plagues our time.

Approach and Sources

My assumption in what follows is that we cannot tell a proper story about the battery cage if we do not consider whether this agricultural technology, in addition to *being made* and *being moved* about (passively, as a thing) also *made* and *moved about* other things (actively, as an agent). This might sound abstract, but my ambition is to make it concrete. For instance, I'll try to account not just for the macro-economic functions of the cage, like the fact that it – as many other agricultural technologies – allows people to move off the land (and away from primary sector work) into cities (toward secondary and tertiary sector work). I'll refer also to the difference it makes to the everyday micro-economics of the farmer and his household. My starting point is thus that, if we are to say anything about the battery cage's path through history, we must begin to take an interest in the diverse entanglements into which it is caught. The battery cage is much more than an artifact made of steel wire; it is a node in a complex network, where memes and meanings merge with dollars and dimes.

A concrete inspiration for this chapter comes from Reviel Netz's book *Barbed Wire* (2004), in which the author shows that barbed wire was not just enabled by a particular constellation of economics, nature, and culture, but that it, in turn, enabled new such constellations. In Netz's story, this seemingly simple technology thus becomes, one could say, an actor in the historical drama. That does not mean that barbed wire develops in self-sufficient isolation; rather, if this technology was an historical actor, it played opposite and alongside a whole cast of other actors. Netz understands barbed wire as a significant part of a larger, complex whole, consisting of unstable and interchanging parts, and he takes it upon himself to unravel the movements of these parts over time.

This chapter can obviously not suffer comparison with Netz's book; for one thing, the story of the battery cage is long and complex, and space here is limited. For the sake of convenience and focus, then, I concentrate on the discourse about cages in the Norwegian poultry association's journal, *Fjørfe*, from the early

1930s through the 1970s – which is the period where the discourse about the battery cages emerges and erupts. I have gone through all issues of the journal from this period. The journal was the main outlet of the community of poultry farmers, and it took upon itself to convey "all the poultry news that's fit to print"; in *Fjørfe*, the content spanned from tips on how to construct feeding troughs to sarcastic commentary on the government's latest agricultural budget. Importantly, the journal was never one-sided; it encompassed a wide array of contrasting viewpoints, in its letters columns as well as in its feature articles.[3] Arguably, it can thus be taken as a workable representation of Norway's poultry farming community.

Hens: From Free-Range Nuisance to Productive Poultry

In Norway, hens had traditionally been considered an inessential part of life on the farm. Up to the turn of the nineteenth century, hens had been quite a common feature on Norwegian farms, but they were typically not considered much of an asset. The amount of input that went into poultry production was miniscule – so small, in fact, that one can hardly think of it as "production" at all. The little surplus that hens provided typically fell to the women on the farm, who would take their eggs to market, and in that way carve out a small measure of economic independence for themselves. Up to the 1900s, however, hens were of no more than marginal importance to life on a Norwegian farm. In fact, many farmers considered these free-range birds a nuisance.[4]

With the establishment of the Norwegian Poultry Association, in the 1880s, this changed – though not very radically at first. Originally a club of urban bourgeois poultry fanciers, the organization's first concern was to upgrade the quality of the chicken material in Norway. The shabby state of Norwegian chickens at the time made them unfit for display at shows, and hence, the association began importing animals from abroad. One unintended consequence of this move was that more people starting seeing economic potential in egg production, which in turned sparked an adjustment of the mandate of the organization towards poultry *farming*. This transition was far from immediate, however, and it would still take several decades before eggs – and later (meat) chickens – became an independent industry (Berger 1997).

Despite the fact that egg production remained a fairly marginal activity during the first decades of the twentieth century, poultry-keeping practices did become more systematic during this period. This development affected the physical structures in which hens were kept, the fodder they were fed, and the care they received. In the first decades of the twentieth century, hens were typically kept in groups no larger than 50 animals, most commonly in quite spacious but fenced-in yards, with a hen house attached. They were fed grain as well as scraps from the household, and, because most hens were kept outdoors at least part of the time, they also searched for their own food. At the same time, the addition of nutritional supplements – not least fish oil – was increasingly seen as an essential part of sound poultry keeping.

The housing system that developed in this period was made up not just of a material structure and of practical routines – but also of norms, including ideas of animal welfare. This assemblage of material structures, practical (including economic) incentives, and normative ideals, fused into a particular notion of poultry keeping that was constantly repeated and consolidated in an emerging genre of this epoch, *the poultry-keeping manual*. From the 1910s through the 1930s, a substantial number of such manuals were published in Norway, written either by amateur farmers or by agricultural scientists, most of whom had their base at Norway's agricultural college. Some popular examples of such manuals were J.K. Selmer's *Praktisk hønsebok* [The Practical Hen Book], P. Wendelborg's *Hønsegården* [The Hen Yard], and *Hønsehold* [Hen Keeping], by an editor of the poultry journal, Leif Svendsen.

The basic presupposition in these manuals was that a happy hen was good for business, "for if the hens do not thrive," as one such manual writer stated, "one cannot expect them to create a surplus" (cited in Berger 1997, 53). Most of those who wrote on poultry keeping in this period were preoccupied with understanding the natural behaviors of the hen. One writer, Nilsen, explained: "In the wild state, hens are most at ease in high grass and on the margins of forests, where they find their food (both animal and vegetable), such as worms, insects, berries and seeds, and therefore they are also better off in domestic settings if they have the greatest possible freedom to roam and to procure a significant portion of their food themselves" (Nilsen, cited in Berger 1997, 54). One writer, Høie, acknowledged that poultry farmers had begun to keep their hens indoors, but concluded: "The sun, the grass, and the movement outdoors will serve a function in the long run" (Høie, cited in Berger 1997, 90). Another, Brandt, wrote: "In summer, the hens must be allowed outdoors in adequately large grass-covered yards, amply lit by the sun … Life outdoors is of great importance if the hens are to thrive" (Brandt, cited in Berger 1997, 90). A third, Engnæs, wrote that one ought to "let them have the chance to move," and added that, "This promotes the growth of muscle and bone" (Engnæs, cited in Berger 1997, 90). None of the writers on poultry in the pre-war era recommended cages. In fact, they did not recommend keeping hens indoors at all, and even those who mentioned this practice, still pleaded that hens ought to be let out at least an hour each afternoon: "This hour of freedom is a consolation to them – they kick around and go searching for many a delicious worm, and around 6 o'clock they often return inside of their own will" (Engnæs, cited in Berger 1997, 91).

The advice in the poultry manuals was largely co-extensive with what one could find in the poultry journal. Here too, the idea was that the poultry farmer's priority should be to keep his hens happy, and that this could be done only by catering to their natural needs and inclinations. In short, the farmer had to make sure the hens had easy access to the outdoors, where they could enjoy plenty of fresh air, sunlight, and the company of conspecifics, not to mention the chance to kick and churn, sandbathe, and roam about in search of food. This overarching "recipe" had to be constantly explicated, refined, and adjusted, however. As anyone who had experience with hens knew, a poultry keeper would inevitably have to confront a series of practical problems and challenges. Consequently, a large part of the material in the journal consisted of a sort of ongoing "trouble-shooting," which aimed first to identify common problems, then to experiment with solutions, and finally transmit to the other readers of the journal the results of those experiments.

It is not far-fetched to think of this pragmatic, experimental approach to poultry keeping as one of *tinkering*. As argued by Mol, Moser and Pols, "Technologies ... do not work or fail in and of themselves. Rather, they depend on care work. On people willing to adapt their tools to a specific situation while adapting the situation to the tools, on and on, endlessly tinkering" (Mol, Moser and Pols 2010, 14–15). This is a rather precise description of what the hen house and yard demanded of the poultry farmer. The basic requirements were laid out in the poultry manuals – indeed, they were crystallized into an ideal. Still, everyone seemed to appreciate that it took effort to make ideals work in practice – hence the need for endless tinkering. The journal performed an interesting function in this regard, for it became a sort of *agora* for individual farmers' tinkering experiences. The discourse in the journal was a sort of collective *literarization*, one could say, of poultry tinkering – it was shared space that allowed the poultry farmers to "tinker together," as it were.

The values that attached to this experimental practice (if not to say, to experimentation itself) are well illustrated by a piece from 1931. The starting point is what it takes to make egg production a viable economic endeavor.

> [A]s with life in general ... it depends on the relevant person's personal abilities, whether a profitable business can be had or not. Consideration, diligence, attentiveness, perseverance, and a sense of order – these are all skills necessary for a good result in any business, and not least in hen keeping. It is *not* achieved simply by having a certain number of animals. One has to *work* with them – and to use one's thinking.
>
> (Bø 1931, 71)

This practical, experimental attentiveness is constantly underlined throughout the entire pre-WWII period, and the journal featured regular columns that exemplify this attitude. For an illustration, consider an installation of a recurrent column, "For beginners," from 1931, by "G.T." [Gerhard Thronsen]: "We ought to remember that hens are among the most habitual of all our domestic animals; for they thrive best when there is no change in their daily way of life." (Thronsen 1931, 9). There are many things to consider, advised the writer, to ensure the laying does not come to a halt. The animals should not be given water, vegetables or grains that are too cold, for "We too have the same experience, that there is no pleasure in going to bed in a cold room after one has eaten cold food and drunk cold milk in the evening." Consequently, feed and drink should be heated to an appropriate temperature. "Activity is another important factor in the daily lives of hens." Cabbage hung from the ceiling and the provision of hay or leaves in which the hens could "shuffle about" was "a good way of making sure the animals get their exercise." In addition, "one needs dry, fresh air." A hen house dense with ammonia was "dangerous." Straw that had condensed on the floor and ground was no good, for "the hen is fully aware that there is no point in working with [it], and then they are left without a cause. And idleness is the mother of vice, etc. And then comes feather-plucking and all sorts of mischief" (ibid.).

The extract is interesting because it displays so vividly the faith in the overarching recipe, but also because it demonstrates the farmer's relation to experimentation. The writer added: "If there is a sudden stop in egg laying, the fault is most likely to be found in us, in the form of some curious desire to experiment – if this brand of fodder is not better than the other one. Nothing is more reprehensible than to change fodders in the middle of the egg laying period." This should not be taken to indicate that poultry farmers were not experimenting; indeed, the pages of the journal are a testament precisely to experimentation. It points, however, to the fact that one believed, in this period, that there was no experimenting with the needs of the hen. Those needs stemmed from the hen's nature, and that was not something one could change as one changes one's shirt. Of course, poultry farmers of the interwar years were not oblivious to the uses of purposive breeding – as mentioned, the initial move towards making a financial prospect of eggs began with the explicit attempt to improve the animal material. But the primary call felt by most farmers was to meet the hens' current needs – the challenge of making *these* hens happy. And that could be done only by catering to their needs, roughly as those were described in the manuals and in the columns of the poultry journal. For our present purposes, the most salient aspect of the prevailing notion of good poultry keeping was the fact that hens needed natural lives, and that this generally speaking excluded tight confinement, i.e. cages.

The big question, of course, is how and why the cage entered into this picture after all? The first thing to note is that it entered *slowly*. Throughout the first decades of the twentieth century, in fact, up until WWII, the system described above was the norm for egg production in Norway. But throughout these decades, the journal brought numerous reports of the use of cages abroad – most notably from the United States, where a battery-cage-based egg production of a certain scale had first emerged in the 1920s. These reports appear not to have been part of any grand campaign either on behalf of, or in opposition to, the battery cage and its introduction to Norway. Rather, most of these articles can be read as a sort of *curiosa*, triggered by the journal's mandate of keeping its readers abreast with recent developments in the world of poultry – whether it was relevant for Norway or not. One finds only very marginal hints to the effect that Norway ought to imitate American developments. Most of those who wrote on the topic emphasized the *exotic* character of this new technology, and typically, the cage was seen as an expression of the American tendency to "go big." Few, at this point, could imagine a scenario where poultry farmers in little Norway would go down the same path. Interestingly, this sentiment persisted for a long time, even despite the increasing air of modernity that descended on the agricultural sector after WWII.

The Bright but Uncertain Future of Eggs

In the post-war years, things turned towards all that was conceived as "modern" in the world of egg production, as well. The factor underlying most sectors during this decade was the post-war sentiment of victory, coupled with the idea that Europe, together with the rest of the free world, was destined for a bright future. Even in the admittedly remote corner of the world that was the Norwegian poultry association's journal, one could sense that the world had now entered the "space age." This became most apparent in advertisements – which, from the 1950s onwards, began to incorporate a space-related aesthetic.

One example of this tendency is a two-page ad in the 1954 November issue of the journal, for "Martens stjernegranulat" – that is, a "star granule" formula marketed by a major supplier of feeds. The ad featured a large image of a hen, surrounded by a series of images of UFOs (the series was supposedly meant to project a sense of movement). At the far end of this series, the UFO(s) seemed to transform into what looks like a trough, in which we find a portion of the mentioned "star granule formula." At the opposite end of the trough is a drawing of an alien, with huge ears and big antennae projecting from its head, who exclaims: "Here you go, a new product in the star class!" The text of the ad underlines the modernity of the product: "The production of 'STAR GRANULA' takes place in a vitamin distillery that lacks an equivalent in all of Europe. The concentrated content of vitamins means that 'STAR GRANULA comes in packs half the size of regular 'Star Fish Oil'." This production, the ad added, "follows upon extensive research work at our own laboratory and at distinguished American and English labs."

Although it might appear to be no more than a curiosity, this type of ad was, in the context of the poultry journal, a novelty of the 1950s – and it arguably signals an emerging faith in a "new era." Admittedly, ads were typically more streamlined than the rest of the content in the journal. If the space age was perhaps enticing, this somewhat fanciful *motif* was – as always – made more complicated by the everyday concerns of the farmer. It had by now become clear to most that Norwegian agriculture had to change – indeed, that modernization and "rationalization" were now almost *fated* – but it was still far from clear what that entailed precisely in the case of egg production. Although some had started having big dreams, most kept "muddling through" in practice.

Down on the farm, a multiplication of production systems had now taken place. There was still the old system, of small flocks kept on outdoor yards, with small hen houses attached. But some had started scaling up, and were now using different versions of a pen-based, indoor system. At the same time, a new debate had arisen concerning the merits of the so-called deep litter system. This was a (mainly) indoor system where the basic idea was to let manure drop on the floor, where it would be mixed with straw that, with time and the hens trampling on top of the mixture, would form a stable, but dynamic floor. Only once a year or so, the "floor" would be removed. If executed diligently, this system reportedly enabled high productivity, since it drastically reduced the need for input of labor.

A farmer who could clean out the hen house once a year rather than once a week, was clearly at an advantage. To succeed with the deep litter system, however, one had to manage certain factors quite strictly, and if one failed to do so, disease and other problems would plague the hen house. Thus, debates about this issue revolved around how well the system performed, when it functioned optimally, but also whether one could get it to function optimally in practice, in the many farms around the country. This goes to show that the need for an experimental, practical attitude remained necessary even as the forms of housing changed. Regardless of the technology, the farmer always had to make it work in practice.

Of course, in the midst of these debates about the merits of the various housing systems, one also discussed the battery cage. Cages had been used in chicken breeding in certain areas since the 1930s, so they were not unknown to Norwegian poultry farmers (Berger 1997). Further, the promises that were usually attached to the cages – a higher productive yield at less input of labor – clearly harmonized with the agricultural ideals of the post-war years, of modernization and rationalization. There was, however, still no widespread conviction in the early 1950s that the battery cage necessarily represented the future for egg production. Rather, this was a period of fairly open and balanced debate on the issue.

One aspect of the battery cage system that seems to have almost uniformly fascinated the poultry community, however, was the massive *scale* it made possible. The concrete implication of this was not necessarily that a similar scale of production ought to be established in Norway. Most argued that this was unrealistic, given the limiting agricultural policies in place. For instance, in 1950, one writer reported from a poultry fair in London, on an exhibition of battery cages. "[I]t was claimed that a battery of this type, with a capacity to hold 10,000 animals, had been ordered to Norway," noted the writer in slight disbelief. "Who the buyer might be, we would not know, and considering our restrictions on concentrates we cannot comprehend how he plans to manage" (N.N. 1950). The underlying logic was that the battery cage required a higher level of investment than did other housing systems, so that a farmer who decided to invest in battery cages had to scale up if he was to have any hope of recouping his investment. Such scaling up in turn required ample access to feed concentrates, however, and given that this was a restricted good in Norway, the writer could not see why a Norwegian farmer would ever make the investment. Given such limitations (another one was the relatively small Norwegian population, coupled with its equally moderate consumption of eggs, not to mention the Norwegian agricultural authorities tepidity when it came to poultry), not everyone considered the battery cage system fit for Norwegian conditions.

Size still mattered, however. The enormous scale that reporters had witnessed in US and the UK – and increasingly in neighboring Sweden – suggested that nothing, in principle, stood in the way of making an *industry* of eggs (and poultry). In Norway, in the 1950s, egg production was still considered no more than a *binæring*, that is, a secondary income – despite the poultry community's persistent efforts to make it something *more* than that. The battery cage systems abroad showed that, not only could poultry be a source of primary agricultural income, it could become a highly *profitable* one at that. The implication of this, however, was once again a certain ambivalence. The American ideal, of making "every farm a factory" (Fitzgerald 2003), was not as apparent in Norway; if there was a sense that agriculture – *all* of agriculture – was and should be modernizing, there was also a keen sense that Norway could not and should not simply do as America had done.

At the same time, when dealers in poultry farming equipment first began advertising battery cages in the 1950s, their arguments tended rather clearly to foretell later developments. Advertisers typically highlighted how the cage expressed the ideal of rationalization. For instance, in an ad from 1954, Gottfred J. Christensen, an Oslo-based supplier of equipment to the industry, used the following slogans as selling points:

> Benefits:
> The hens stay healthier and thrive
> The animals always receive adequate feed and water
> The production of eggs is increased
> Animal handling takes less work
> No more feed thieves
> Battery cages are a profitable investment

In another ad, the same dealer announced that their battery cages, which were adapted either for raising chicks or for egg-layers, all could be bought "with or without mechanical cleaning, fully automatic feeding and watering systems." The main Norwegian producer of poultry machinery and other equipment, Ole Hauger, placed an ad in 1954, which stated that demand had been so great that they could foresee not being able to meet all requests. The message to poultry farmers was that an order would have to be placed *today*. Although not necessarily more than a salesman's pitch, this does tell us something about the way things were going, or at any rate, the atmosphere that certain actors were now trying to create. More concretely, these actors were trying to create an atmosphere of inevitable progress, as if mechanization was the natural and necessary route to the future. Or as Hauger wrote, in an ad for a starter cage: "Hauger's chicken cages with manure-remover-belt is the … dream of all chicken producers." Advertising is the manufacture of desire, as we know, and at this point in time, advertisers were telling the poultry farmer that what he should desire was to have his manual work taken over by a machine.

A general shift had now begun to make its way into the discourse, towards the assumption that the future would be one of larger scale and less input. This entailed a certain adjustment of the pre-war ideal, not least a move away from the notion that hens had to lead *natural* lives. For instance, in May of 1951, a piece in the journal referred to an article in the German poultry journal, *Kleintier-Welt*, which read:

> One thinks that laying hens left singly in battery cages are degraded to dead machines. The animals stand on wire floors, in wire cages, and are fed and watered automatically. The eggs tumble down into a collecting tray – and so goes day after day in this hen's life. One feels sorry for hens that cannot come out in the free to run around and chuckle.
>
> (N.N. 1951)

This, the German article continued, was "against nature," and would undoubtedly lead to higher mortality. In response, the Norwegian editor of the journal remarked, that, "We would like to note that the battery cage facilities we have seen have always been completely first rate … If the animals are not happy and are not getting enough fodder, water, vitamins, and not least, air, the first sign of this would be a reduction in egg production and an increase in mortality. But the opposite is the case." The editor was not defending the practice on the ground in Norway, since hardly any Norwegian poultry farmers were using battery cages for laying hens at this point. Rather, he can be seen to express the more general attitude of the Norwegian poultry community at the time, where the – ambivalent, and therefore open – response was basically to keep all options open, since one did not know for sure what the future would bring. The prospect of a transition to battery cages in Norway had to be kept open, but only *alongside* other – at this point equally viable – options.

Cage Debates

By the early 1960s, the relative merits of the deep litter system (and other forms of housing) had been downgraded, and those of the battery cage correspondingly upgraded. The path to the future had begun to crystallize. By the end of the decade, that path had lead to a situation where most Norwegian poultry farmers operated with battery cage facilities, and by the end of the next decade, almost all Norwegian egg producers housed their hens in cages. According to Kjersti Berger, a historian of the poultry industry in Norway, this transition fundamentally changed the prevailing notion of what constituted a good life for an egg-laying hen. "Earlier," she writes, "rational management was synonymous with sensible management, and sensibly managed was a poultry system that accounted for the natural behavioral needs of the hens. Now rational management came to be synonymous with less time-consuming management" (Berger 1997, 93).

This should not be taken to mean that "less time-consuming management" had previously *not* been part of the farmer's concerns. Indeed, the main selling point of the deep litter system was precisely that it saved the farmer labor, and thus time. And a large part of the constant tinkering of the pre-war poultry farmer, as described above, consisted in developing practices and techniques that could save the farmer time and effort. The poultry farmer was constantly trying to get more (or at any rate, *as much*) with less. The implication of the transition was rather that, with the arrival of the cage, one talked less about the concrete, natural needs of the hens. It was not so much that these needs were contested; rather, one simply stopped talking about them, and focused instead on the hens' output, as a supreme measure of their welfare (see Bjørkdahl 2015).

This was another central implication of the introduction of cages, according to Berger, that the poultry community – now aided by the scientific environment at Norway's agricultural college as well as its equivalents abroad – decoupled the notion of natural lives from that of productivity. It was not true, one now realized, that hens would produce only if and when they had access to sun, grass, fresh air, and plenty of space. Quite simply, *a hen did not have to live a natural life in order to lay eggs*. Consequently, the equation that had been established in the pre-war poultry journals was turned on its head. One no longer assumed that a hen needed to live a natural life in order to thrive, and that it needed to thrive to lay eggs. Instead, one stated that caged hens laid as many eggs as free range ones, and hence, they *must* be thriving. In 1960, for instance, the leader of Norwegian Egg Central stated that: "Many believe [battery cages] are unnatural and provide little 'pleasure' for the hens. But the truth is that the hens both lay eggs and appear to thrive in cages like these" (cited in Berger 1997, 111). What largely dropped out of this equation, of course, was the question of natural behaviors.

An intensification of the discourse about the battery cage begins in 1963–1964. The most striking feature, in the numerous pieces on the issue from these years, is the persisting presence of arguments that complicate or downright advise against the introduction of battery cages – arguments that were put forth by farmers, authorities, and agricultural scientists, though, naturally, *not* by producers of farming equipment. Of course, many others came to a more positive conclusion – and a substantial number arrived at a more ambivalent assessment. In April 1964, one writer, Birger Ukkelberg, wrote a piece about the prospects of "big poultry and cage systems" in Norway. His article is notable for the way it illustrates the relevance, at this point, of weighing the *pro et contra*.

> Now and again we hear plans about some large poultry farm about to happen. The plans may be left at the drawing board, or they can progress considerably. They can be based either on cage system or pen system, but predominantly cage systems, because it allows room for more animals. The investments are considerable, but according to the plans, there will be so many animals on which to share them, that the venture will in any case be profitable. Things will be so mechanized and easy that one man can care for several thousand hens. As time goes by, Norwegian poultry folk will have to make up their minds about the egg factories, if they have not done so already.
> (Ukkelberg 1964, 51–52)

Ukkelberg did not question the farm household economics of the battery system. And in general, no one challenged the claim that the battery system entailed an easing of the farmer's workload. But many commentators, including Ukkelberg, noted that household economics was not the only relevant perspective. Indeed, one had – as was also the tradition in the journal – to see egg production in a context of national (perhaps even international) agricultural production. Battery cages might mean a more comfortable work situation for the egg producers but, as Ukkelberg put it, "If some [farmers] might benefit, it is nevertheless doubtful whether it is rational that other hen houses and farm buildings here and there perhaps are left empty for precisely the same reason" (ibid.). The underlying logic was that the battery cage was eminently efficient – "rational" – when seen on a farm/household level, but that it at the same time challenged the balance of Norway's agricultural production. The traditional hindrance had been the rationing of fodder, which at this point had been lifted. Now, in 1964, what curbed the enthusiasm of certain members of the poultry community was that the current consumption of eggs could not justify any further rationalization of egg production.

The ghost hovering in the background of Ukkelberg's analysis, of course, was the prospect of *overproduction*. Referring to a professor at Norway's agricultural college, Fridtjov Sannan, Ukkelberg noted that the expert had advised caution when it came to the question of cages, claiming that the "battery cage system was not a short cut to riches." "It is not my impression," wrote Ukkelberg, "that the public institutions for training and guidance have been particularly firm in their recommendation of battery cages." And why had they not? "When the market is now telling us that we don't need this increase in capacity, there is a danger that the investments in cages are wasted" (ibid.). Whether he wanted it or not, each individual farmer was tied to the overall market, Ukkelberg pointed out: "If a producer has to halt his production of eggs a while or switch to another type of production, this is so much harder if he has made an investment in cages" (ibid.). Consequently, not even on the farm/household level could one conclude that the battery cages were "rational."

The journal also reported on other complications that resulted from the cage system, among them animal welfare issues. In an article from 1964, one writer warned of a new ailment that befell battery hens, called "cage fatigue." "The main motive [behind the introduction of cages] has been a greater profit," noted the writer, so "that the work is made easier, that many routines are mechanized, and that more hens are housed per square meter of floor" (O.T. 1964, 107). But this prospect of profitable simplicity was threatened by "an inconvenience called *cage fatigue*," which the writer described as a particular "weakness of the limbs." "The hens have trouble standing up and moving around. They support themselves with their wings and in the worst cases, lie down on their sides. Investigations of such hens have often revealed fractures in the scapula and the clavicles." Cage fatigue was something that in particular befell hens with "a high laying percentage," the writer noted, before he added that, "animals sick with the disease recover quickly when they are placed on a floor or on deep litter" (ibid.). If this suggests that the discourse on battery cages was not blind to consequences for animal welfare, the existence of this disease did not weigh heavily enough to lead the writer to advise against cages. The precise cause of the disease (i.e. more precise than the caging of hens) was not yet known, the writer noted, as he went on to supply a list of six tips for producers whose hens struggled with cage fatigue, among them recommendations for fodder and additives, choosing hens bred for cages, and allowing hens that suffered from the condition onto a floor to recover.

We can take this as an illustration of how the tinkering discourse, which had been so ubiquitous in the journal ever since the 1930s, now moved into another frame. The tinkering impulse was much the same, as each individual farmer still had to confront basically the same sorts of questions: How can I make sure my hens are healthy and productive, while caring for myself (my health, my work load, my family, and finances)? However, the farmer now increasingly found himself tinkering with the features of an industrialized farm operation. Now, he had to experiment with fodder, with nutritional supplements, with machines (mechanical watering pipes, feeding trays, manure-removal belts, etc.), and – not least – with medicines, notably antibiotics. Incidentally, the latter had proven to have little effect on cage fatigue. "Supplements of minerals, antibiotics, and hormones have been without effect," noted the writer, and added that "it has been claimed that cage fatigue is caused by *stress factors* that originate in the battery hens' lack of movement" (ibid.).

The shift from a manual to a mechanical – from an inventive to an industrialized – tinkering discourse was neither immediate nor clear-cut. Indeed, for quite a few years more, debates in the journal would include pieces that advised *against* the transition to battery cages, not just on the grounds of national agricultural economics, but for reasons of animal welfare. The air of reservation and ambivalence that still framed the discussion was about to be marginalized, however. Starting in 1964, the journal brought an increasing number of articles that argued for the new technology. In the August 1964 issue, for example, the journal reported from Hovelsrud, where "each part of the farm is operated intensively and rationally" (Dahl 1964, 133). Battery cage egg production was one of those parts, and the family, Olssøn, had recently converted from a pen system to a self-built battery cage system, modeled on what the elder Olssøn had witnessed abroad. The article consisted of an extended interview with Olssøn, accompanied by images from the farm, as well as captions commenting on the type of wire used for the cages, the construction of the feeding trays, water pipes, etc. Asked what he wanted to convey to the journal's readers about battery cages, Olssøn replied:

> We believe battery cages are the best and most economical aid that technical development and modern science has provided egg producers. You can have a higher level of production with less input of fodder, better hygiene and thereby higher level of thriving hens, less discard and lower mortality, longer and more efficient laying period, clean eggs, and a minimal input of labor.
> (Dahl 1964, 133)

A more enthusiastic case for the benefits of the battery cage system could hardly have been made, and this was the testament of a farmer who had quite recently converted from the pen system – and who was apparently in a position to make pronouncements on both. Olssøn was now a cage enthusiast – so much so, in fact, that he had placed an ad on the following page, announcing that he was producing cages for sale: "The cages are solid, they are simple, they are Norwegian, and they can be bought at the lowest price available on the Norwegian market today." Importantly, the writer noted in his conclusion to the piece:

> On my own account, I will add that I got the very best impression from the cage system at Hovelsrud farm. The animals seemed very happy and their achievements appeared to be top notch. The eggs looked clean. Cracked eggs were not seen during our visit. As far as this writer is concerned, it looked as though Olssøn has solved the construction of battery cages very favorably. (Dahl 1964, 135)

This type of case was developed, standardized, and repeated during the rest of the decade. Things came to a head, one might say, in 1965, when the question was made explicit in a series of articles under the heading, "Ought hens to be placed in cages or in pens?" The debate started with a editorial, by the editor Sigurd Bjørnstad, where he wrote that, "What will become the housing system of the future in egg production, we do not yet know – maybe it will be cages, or maybe pens, or as now, both" (Bjørnstad 1965a, 93). The downside of putting hens on the floor, in pens, wrote the editor, were by now well-known. "This is not the case with cages. Right now, this is fashion. It is 'news,' and when combined with the advertisements, this easily causes us to see the benefits and overlook the disadvantages" (ibid.). A definite answer would not be forthcoming in at least 15–20 years, wrote the editor, and in order to avoid "bad investments" he thought it best to "advise caution, which is also what sales people and [the authorities'] consultants have done" (ibid.). Bjørnstad went on to warn his readers that many of those now selling battery cage systems were the same who had done the "experiments" which supposedly should cause one to have confidence in the battery cage system. As long as this was the case, he noted, one could only expect a "one-sided presentation of convincing results of experiments in favor of cages" (ibid.). Between the lines, the editor was referring to Olssøn and a lecture the latter had given, where he had claimed that his very satisfactory "experiments" with battery cages had convinced him that Norway should make a large-scale transition to this production system. In an attempt to balance this state of affairs, the editor noted, the journal would in the present issue bring an article by the professor at Norway's Agricultural College, Fridtjov Sannan – a longer background article that aimed to provide a scientific *status presens* of the merits of the battery cage system relative to the pen system.

The editor's text appears to have infuriated Olssøn. Writing in response to the editorial, Olssøn explained that he had not claimed to have done real, scientific "experiments"; he simply wanted to convey his experiences after three years of using (and building) battery cages. He made sure to note that, "After having produced eggs in battery cage systems for three years, we are, like many international producers, convinced that this housing system is the supremely best one economically speaking. Production of battery cages is a very small part of our operation" (Olssøn 1965, 129). Olssøn rejected the notion that his lecture was made as a sales effort, but acknowledged that "problems are to be debated" (ibid.)

He concluded, however, by urging the editor to visit battery farms around the country, so that he, "to the benefit of all readers of the journal, can take part of the impressive development this industry is now in" (ibid.). To this, the editor replied laconically: "The results of Mr. Olssøn's own 'experiments,' to which he refers in his piece, strictly speaking demonstrate only that he has acquired a better 'grip' on cage housing than on floor housing in his own egg production. Meanwhile, other producers have had negative experiences with caged hens. What is to be preferred has therefore to be decided in each case after critical reflection and careful calculation" (Bjørnstad 1965b, 129).

This exchange can illustrate that there was no very clear-cut point at which the case for cages was definitively won; rather, the actual (and growing) presence of battery cages in farms all around Norway seems only gradually to have become a silent premise of the debate. Naturally, when a substantial number of Norwegian egg producers had switched to the battery cage system, the point many had made earlier, that this system came with plenty of problems and deficiencies, was much harder to make. What is interesting, however, is that precisely when the cage becomes a premise for the discourse on egg production, the public awakes to what has been going on – behind their backs, as it were. Thus, as we approach the end of the 1960s and the beginning of the 1970s, the discourse in the poultry journal shifts from negotiating between past, present, and future within the poultry community towards an effort to defend that community from attacks from the outside.

In Defense of the Cage

By the early 1970s, battery cages had become the norm for Norway's egg producers. Whether or not to install cages was no longer the issue; now one discussed *which type* of cage system to install. This does not mean that the debate about battery cages disappeared. In fact, it only intensified further in this decade, although it also *changed* significantly: Whereas earlier, the debate about cages had divided the community of egg producers into different camps, now this community increasingly found itself defending the battery cage against attacks from the outside – not least from animal protectionists who, with the new Animal Protection Act in hand, argued that the cage prevented hens from living out natural behaviors.[5]

The 1970s saw the emergence of a revitalized animal protectionist movement in Norway, and with it came a more vehement critique of the battery cage. Ruth Harrison's *Animal Machines* was translated into Norwegian in 1965, and it formed the backdrop for much of this critique. The translation incidentally featured a foreword written by the veterinarian Inga-Johanne Jebsen Haave, who was a profiled veterinarian/animal-protection activist at the time. Other important parts of the protectionist landscape were the already mentioned Kåre Knutsen and the Campaign against Caged Hens.

Even before his book came out, in 1974, Knutsen had stirred things up, not least with an op-ed in 1972, for *Arbeiderbladet*, entitled "Sun Eggs – Modern

Animal Abuse." The poultry journal found it imperative to respond to Knutsen's onslaught – and it did so with a series of articles, including a feature by Fridtjov Sannan, the professor at the agricultural college, who now wrote in his capacity as "state consultant for small agricultural business [including poultry]." Sannan's text was a "commentary," a 7-page hybrid cross between an opinion piece and a scientific report, where his explicit aim was to counteract the impression left by Knutsen's article. The journal promised a report that would be "a useful corrective to those who now want to attack 'modern' poultry farming" (Sannan 1972, 134). This is quite a significant statement, considering that the same Sannan only a few years previously had advised caution to those who promoted the battery cage too enthusiastically.

Sannan began by establishing that, in egg production as in other production industries, prices of the means of production had increased more dramatically than prices of products, i.e. that the farmer was *not* reaping larger profits than before. In fact, it was the other way around, and this meant that the whole sector was under some pressure. Sannan was straightforward about the economic realities: "To ensure supply for our national demand for fresh eggs would be a hopeless task, economically speaking, for Norwegian poultry industry if the conditions in this industry were so static as to stick with the housing forms, production methods, and industrial structure common in the 1930s and 40s" (op cit. 135). This, added Sannan, was "part of the explanation for today's egg production, which despite the unfortunate price development, stands as a vibrant, competitive business" (ibid.). This was due to a number of factors, among which was not least, "Rationalization of work through more expedient construction of houses, interiors, and equipment" (ibid.). Since Sannan had made his previous intervention, demand for eggs had gone up. One no longer used *low* demand as an argument *against* battery cages; instead, one used *high* demand as an argument *for* them.

More to the point of animal abuse, Sannan wrote: "The fact is that it has never before been less apposite to use the term animal abuse about Norwegian poultry farming than it is today. Poultry farmers have always understood that to achieve an *economic* production, the hens must enjoy as good conditions as possible" (ibid.). The question of "cages or no cages for laying hens is, to many people, more a question of sentiment, habitual thinking, and personal considerations than of actual information and sound argument" (ibid.), noted Sannan, and in response to this situation, his article took the form of a kind of fact sheet for battery-cage-based egg production. In his review, he did not underplay completely the fact that housing hens in cages came with certain challenges, but his attitude throughout was one of a willingness to tinker – to fix what problems might be.

Sannan firmly rejected proposals for a law that prescribed that all hens should have access to an outdoor area. Such proposals, he said, "seem quite unrealistic" (op cit. 140). The trouble, again, was the needed scale of the production, seen against the practicality of the housing system. "In order for such yards not to become immense sources of infection, one would have to move to a new, clean yard with only a few years in between. How one, in the long run, could get enough yards with direct access to the hen house, is quite the mystery" (ibid.). Besides, he added, "after we learned to make real concentrates it is no longer necessary to rely on the nutrients the hens would find in a yard more or less covered in grass" (ibid.). And what about Knutsen's complaint about the brand name, *Sun Eggs*? Sannan gave short thrift: "As most other foods, eggs consist for the most part of the energy from the sun, accumulated and transformed. So the designation 'sun' can reasonably be used for eggs as well as for many other products" (op cit. 141).

As mentioned, the discourse in the poultry community had begun to quite systematically *circumvent* the traditional notion of good poultry keeping – the one that emphasized the hens' natural needs – and, in his own way, Sannan contributed to this move. Although he did also deal directly with animal welfare, the overall effect of his argument was to marginalize such concerns. The central feature of egg production was now – finally – *scale*: The battery cage is eminently productive, and no other production system can meet the demand, was his basic retort to critics. Throughout the 1970s, this line of argument – one could say "defense," since the community was now increasingly under attack from animal protectionist critics – became standard.

In 1973, for example, the journal reported from a big poultry symposium in Uppsala, Sweden. One N.O. Lindgren, a veterinarian, stated that:

> Intensive poultry keeping is characterized by large units within the same walls, under the same roof. From a veterinary standpoint, the congestion of many animals to a large crowd has usually been considered incompatible with good health. These views have now generally been refuted.
>
> (Lindgren 1973, 37)

One might conclude that Lindgren thought hens did not really *need* the factors called for by the traditional notion of good poultry keeping, like fresh air, sunlight, the chance to move about and kick in the ground and search for food, etc. That was strictly speaking not the case, however. Speaking on behalf of veterinary medicine, Lindgren, like many others, subtly sidestepped the older notion, and replaced it with a new one. Giving the issue of natural behaviors no attention at all, the central objective was now to establish absence of disease as a precondition of "yield." "A good hygienic standard, recruitment on the principle of all-in, all-out, effective vaccination programs, and where possible, an absolute absence of infections" (ibid.) – these were now the preconditions of a sound egg production, and they were, interestingly, all of a nature that called for veterinary expertise. Naturally, this was no coincidence. In an interview in the same issue, Lindgren elaborated:

> Veterinary medicine has no particular objections to the intensification of poultry keeping. For veterinary medicine too, the primary goal is to achieve the greatest possible profit from our animal production. Veterinary medicine itself has given valuable contributions to this development and has to a large

> extent made the intensive poultry production of today possible. I would not be exaggerating if I claimed that poultry production, more than other forms of animal agriculture, has put the recommendations of veterinary medicine into practice. The principles of veterinary medicine are now so well incorporated that they are viewed as obvious, routine.
>
> (Lindgren 1973, 36)

Lindgren added that "the design norms for the hens' environment that make possible the optimal use of the animals' individual capacity to produce and survive, are fully satisfactory," meaning that the current norms for cages and equipment were good enough. "Where the requirements for this 'production optimum' are satisfied, there are, from the standpoint of animal welfare, no reasonable additional requirements," he said (Lindgren 1973, 36). In other words, Lindgren was saying that, as long as one supplied environments that allowed the hen to be maximally productive, there were no additional "needs" the hen might have that we ought take care of. Or, as a header in the article put it, *High production is a sign of a good environment*.

This rather more "hard-nosed" approach to poultry keeping was increasingly contrasted to the supposedly softer stance of the animal protectionists. As an illustration, in 1976, the poultry association organized a panel on "Poultry Farming, Animal Welfare, and the Animal Protection Act," from which the journal printed a report. "In the last few months," noted the journal, "our daily press has printed a long line of emotional attacks on caged hen production, and this, the first debate where one aimed at rationally trying to deal with the various issues, was welcomed with great interest" (N.N. 1976, 215). The wording was not a coincidence. An emphasis on the critics' *irrationality* and "emotional attacks" was to become a standard feature of the poultry community's response to criticism throughout the 1970s. One of the panelists at the symposium, Finn Gjesdal, a lecturer at the Veterinary Institute, noted that, "One works hard at such philosophical thoughts [concerning the suffering of other beings], but one must not take one's philosophizing so far in the direction of ethical extremism or desire for compassion that one expels our biologically self-maintaining reason" (ibid.).

Argumentatively speaking, this response was quite disingenuous. The critics were of course no less "rational" than the various members of the poultry farming community, nor were they more emotional. They simply based their evaluation of the current way of producing eggs, and keeping hens, on another – older – notion of what "good poultry keeping" entailed. Whereas the critics – in line with traditional beliefs and, arguably, the Animal Protection Act – assumed the natural behaviors of the hen to be a significant factor, poultry farmers had begun to marginalize this concern only to highlight more "technical" features, most notably yield. Thus, in the 1970s, we can see not only how the discourse on the cage changes within the poultry community, in an attempt to prepare a defense from outside attacks; one can see also the great distance between producers and consumers of eggs – if not to say, of agricultural products.

Over time, the poultry community had been moving away from the notion that free movement was a prerequisite for good poultry keeping. It was not so much that one no longer acknowledged that it might, in isolation, be a good thing. Rather, the poultry community's position was based on the fact that, given certain economic realities – which had not been present in the same way in traditional, very small-scale egg production – caged hens were as productive, if not even more productive, than those in other housing systems. And the cage, of course, offered many additional benefits to the farmer. It was hygienic, it demanded less input of labor, and it opened the prospect of more animals, and thus more income. Importantly, the cages also promised more control – hens in cages could be monitored much more closely than hens on a floor, not to say outdoors. Also, hens in cages were not as free to enact a pecking order, what could of course also be presented as an animal welfare measure.

What had happened, then, was not so much that the traditional notion of good poultry keeping was discredited. No one was saying that hens are not supposed to move around, and no one was saying that these are not natural behaviors for a hen. Instead, what happened was that – as egg production, *thanks to* the battery cage, became a viable form of agricultural production – the economic aspect became the dominant one. One no longer started with the question, "How can the hen thrive, so that it lays plenty of eggs?" Rather, having introduced the cages, the question was now, "How could hens as productive as these be anything but well?" The question of natural behaviors largely dropped out. The field of argument had thus shifted. Earlier, one had assumed that the farmer had *first* to supply surroundings that spoke to the natural inclinations of the hen, only *then* to expect the hen to do her job. Now, one assumed that, if experiments could prove that caged (hence "unnatural") hens were as productive, that was all one needed. The imperative to provide an environment that allowed natural behaviors was sidestepped.

Conclusion

I have here sought to establish three inter-related theses: 1) the introduction of the battery cage to Norway was far more complex than what is allowed by the popular narrative, where the cage is often portrayed as a natural and streamlined part of the brutal apparatus of industrial animal agriculture; 2) gradually, the poultry community began to take the cage as a premise, and the old tinkering discourse shifted from an inventive to an industrial sphere; at the same time, the community began using another – more "hard-nosed" – stance when responding to the increasingly frequent attacks from the outside; 3) when this happened, the result was a series of instances that displayed the increasing distance between the agricultural community and the general public.

We have thus ended up where we started, as it were, at the claim that the egg producers willfully *misrepresent* how eggs are produced. The infamous rustic egg carton was one example; another is the meat and dairy cooperative's consistent use of euphemisms when referring to the cages that – despite the EU regulation –

are still in use. Given the history I have tried to tell, what can be said, in conclusion, of this tendency?

From the perspective of the general consumer, or even more so, from the perspective of the animal activist, the industry is using disingenuous advertising; what the industry presents is too often far from reality. Furthermore, that reality often falls far short of what a consumer can rightfully expect. The consumer does not want to be made an accomplice of animal abuse. Some would even say that the industry is at odds with the law, which ordains that we are to provide animals with the chance to live out natural behaviors. From the perspective of the farmer and industry, however, these objections fail to take into account the reasons why the battery cage finally became so ubiquitous. In short, the cage offers an unprecedented output at minimal input (of labor). And because this is so, the battery cage has been one of several key components of a larger agricultural shift, where we need – and expect – the same, if not even an increased, yield from our agricultural sector, while we expect – and accept – that a steadily lower number of people are employed in agricultural work. To make it very simple, we live in a society where almost no one wants to farm, but where we have become accustomed to a steady stream of safe, plentiful, and cheap food. And this is an equation that does not add up.

The history of the battery cage explains this process quite well, since the central nexus, as we have seen, is where input of labor (time-efficiency of the housing system) meets output. The underlying motif of rationalization, which had its sociological base in the shift from a population employed in the primary sector to one employed in industry – and later, service sector – jobs, meant that each individual poultry farmer, as well as the community as such, had to look for ways to become more efficient. And this was precisely what the battery cage provided. Once the individual farmer had shouldered the cost of a battery cage facility, he could have hopes of producing an unprecedented number of eggs with similarly unprecedentedly low levels of labor. The transition was not without risk, because it did mean that a great many poultry farmers were bankrupted. But in the larger scheme of things, this was simply what was meant by "modernization." The important thing in this context, however, is that, from the perspective of the farmer, this modernization conditions how eggs can be produced. In a modern world, one simply cannot produce the numbers of eggs that consumers apparently expect, at the prices they expect, without (something like) the battery cage. The only alternative is for more people to acquire hens.

At the same time, these opposing perspectives do not quite explain the tendency of producers of meat, milk, and eggs to market their products with representations of how farming looked (roughly) 100 years ago. It is as if the producers assume that consumers would be less inclined to buy if they knew how the food was actually produced. This assumption is probably quite accurate, and there are some very natural reasons why this is so. In their everyday lives, today's consumers are removed not just from the production of meat, milk, and eggs, but also from the rhetorical justification of the same. The agricultural sector is to a large extent a closed rhetorical circuit, as it were. The discourse on the battery cage is a perfect example; it was open and balanced and experimental up to the point where critics attacked the poultry community, and then the black box was closed – all disagreements laid to rest.

This has left us in a situation of miscommunication over agricultural products. Actors like the Egg Centrals cannot market their products using realistic images from contemporary egg production, because the general public has not been accustomed to and conditioned by this discourse – they have not, for instance, been frequent readers of *Fjørfe*. From this conclusion, one can venture further speculations, namely that this particular case exemplifies a more general tendency when it comes to consumers' relation to agricultural production. Today, and in part precisely *because* of the tremendous economic-technological success of industrial agricultural technologies, most people have no everyday experience from farming. They are thus not conditioned – as those invested in the agricultural domain are – to thinking of animals as "units," to counting them in numbers (or even in *weight*), or to talking about them as machines, as transformers of feed into human food. Hence, as the advertisers have clearly realized, agricultural production must remain somewhat of a mystery to the consumers, while the output of that production – like the "Sun Eggs" on the controversial carton – is best sold with appeals that recall older farming practices, including those romantic images of poultry farming – where "[e]verything smells of sunshine and health," and where hens walk "leisurely about, nibbling on grain and straw."

Acknowledgements

This chapter arises from research that was sponsored in part by the Norwegian Animal Protection Fund. Small parts of this chapter appear, in slightly altered form, also in Bjørkdahl (2015).

I thank the Norwegian Poultry Association for their cooperation, notably for facilitating access to their archives of historical issues of their journal.

Notes

1 All quotes in Norwegian have been translated by the author.

2 The EU directive simply establishes a set of requirements for the battery cages, the most important of which is the available space per hen, the availability of nest, litter, and a place to perch. There is, in effect, no ban on cages, but rather, a regulation that enacts a transition from so-called "bare" to so-called "enriched" cages, see EU (1999).

3 If the journal provides a rich and handy source, there is obviously much one cannot access by restricting oneself in this way. For that reason, I draw on on Kjersti Berger's review of the history of the industry in Norway (Berger 1997) for context, as well as my own work on the newspaper debates about the battery cage (Bjørkdahl 2015).

4 As late as 1931, a writer notes that, "Today, no one is offended if elderly gents speak of poultry and all that relates to it. In many places, there is even quite some interest in these winged creatures that, no more than 15–20 years ago, were seen as a great annoyance on most farms, at least for the husband on the farm" (Overaae 1931, 20).

5 Article 2 of The Animal Protection Act, from 1974, stated that, "The instincts and natural needs of the animal should be taken into consideration, so that it does not suffer needlessly." (see www.lovdata.no).

References

Appleby, M. (2003) "The European Union Ban on Conventional Cages for Laying Hens: History and Prospects," *Journal of Applied Animal Welfare Science*, 6(2): 103–121.

Berger, K. (1997) *Fri som fuglen? Modernisering av hønseholdet i Norge, 1880–1975*. Thesis. Oslo: University of Oslo.

Bjørkdahl, K. (2015) "Caged Welfare: Evading the Good Life for Egg-Laying Hens," in K.L. Syse and M.L. Mueller (eds) *Sustainable Consumption and the Good Life: Interdisciplinary Perspectives*. London: Routledge, pp. 204–223.

Bjørnstad, S. (1965a) "N.T. [editorial]," *Fjørfe*, 4(82): 93–94.

Bjørnstad, S. (1965b) "N.T. [response to Olssøn]," *Fjørfe*, 5(82): 129–130.

Bø, E. (1931) "Hønseholdets lønnsomhet," *Tidsskrift for fjærfæavl [Fjørfe]*, *5*(48): 71.

Bøhm-Pedersen, K. (1996) "Høne-kampen begynte i dag," in *Aftenposten*, 16 September, p. 6.

Dahl, Ø.T. (1964) "Besøk på Hovelsrud gård," *Fjørfe*, 8(81): 133–135.

Davis, K. (2009) *Prisoned Chickens, Poisoned Eggs: An Inside Look at the Modern Poultry Industry* (rev. edn). Summertown, TN: Book Publishing Company.

EU (1999) "Council Directive 1999/74/EC," online at: http://eur-lex.europa.eu/legal-content/EN/TXT/?qid=1435151498494&uri=CELEX:31999L0074.

Fitzgerald, D. (2003) *Every Farm a Factory: The Industrial Ideal in American Agriculture*. New Haven, CT: Yale University Press.

Gleditsch, U. (1976) "Aksjon mot burhøns," *Aftenposten*, 2 February, p. 9.

Harrison, R. (1964) *Animal Machines*. London: Vincent Stuart.

Knutsen, K. (1974) *Dyrenes rettigheter: Ulovlig og lovlig dyreplageri i Norge* [The Rights of Animals]. Oslo: Dreyer.

Lindgren, N.O. (1973) "Veterinære synspunkter på intensiv fjørfeproduksjon," *Fjørfe* 2(90): 35–36.

Mol, A., Moser, I. and Pols, J. (2010) "Care: Putting Practice into Theory," in A. Mol, I. Moser and J. Pols (eds) *Care in Practice: On Tinkering in Clinics, Homes and Farms*. Bielefeld: Transcript Verlag, pp. 7–25.

Netz, R. (2004) *Barbed Wire: An Ecology of Modernity*. Middletown, CT: Wesleyan University Press.

N.N. (1950) "Teknikk. Moderne batteribur," *Fjørfe*, (4(67): 86.

N.N. (1951) Batterisystemet (unaturlig)," *Fjørfe*, 5(68): 117.

N.N. (1976) "Fjørfehold, dyrevern og dyrevernloven," *Fjørfe*, 8(93): 215.

N.N. (1981) "Punktum for burhøns-debatt," *Aftenposten*, 8 April, p. 23.

N.N. (2011) "Alle burhøns er løslatt," *Dagbladet*, 30 December, p. 13.

Nortura (2011) "Burhøner – snart en saga blott!," 5 August, online. https://medlem.nortura.no/alle-nyheter/burhoner-snart-en-saga-blott-article29380-11937.html.

Olssøn, N. (1965) "Leserinnlegg: Skal høner gå i bur eller binge," *Fjørfe*, 5(82): 129.

O.T. (1964) "Burtretthet hos verpehøner," *Fjørfe*, 6–7(81): 107–108.

Overaae, H. (1931) "Eggsamvirket i Norge," *Fjørfe*, (48): 20.

Sannan, F. (1972) "Moderne dyreplageri – solegg," *Fjørfe*, 5(89): 134–141.

Singer, P. (1975) *Animal Liberation: A New Ethics for Our Treatment of Animals*. New York: New York Review of Books/Random House.

Thronsen, G. (1931) "For nybegynnere" *Fjørfe* 1(48): 9.

Ukkelberg, B. (1964) "Litt om storhønseri og burdrift," *Fjørfe*, 4(81): 51–52.

5 Back to Nature!

Rehabilitating Danish Research Monkeys

Lene Koch and Mette N. Svendsen

In the spring of 2004, twelve Capuchin monkeys were transferred from a psychiatric hospital laboratory to a small private zoo, in order to be rehabilitated. The monkeys were transferred at a time when the conditions in Denmark had made it undesirable to continue nonhuman primate research. The moral and regulatory landscape had changed and the housing arrangement at the lab would become illegal by 2007. The lab was part of the Skt. Hans Hospital (SHH) near Copenhagen; the research agenda of the psychiatric community was in flux and the scientific need for these specific monkeys was dwindling. Furthermore the public approach to nonhuman primate research was highly critical after an intensive media campaign led by animal welfare organizations to save the SHH monkeys from routine killing. After a prolonged and complicated negotiating process it was decided to move the monkeys from their narrow, individual steel cages in the hospital's basement and transfer them to a zoo with specially designed, spacious collective facilities with outdoor access (see Koch and Svendsen 2015, for a more detailed study of the transfer process).

Several actors were involved in making the transfer possible. Danish animal welfare organizations were instrumental in creating a strongly critical media focus on the animal's situation; the psychiatric hospital was ready to close down the lab for many reasons, among which was the wish to avoid unwelcome public scrutiny; zoo-based primatologists expected that the monkeys could be used as a genetic breeding resource; also the pharmaceutical industry, which had been funding substantial parts of the research, found the increasing public criticism of the conditions of laboratory animals worrisome, and saw a possibility in experimenting with new housing forms for primates in order to continue research under new morally acceptable circumstances. In spite of their different interests and viewpoints, these actors managed to constitute an informal alliance, making the transfer possible – financially, morally and practically. During the period between 1998 and 2003, difficult negotiations between these actors took place. Some considered the monkeys' rehabilitation an end in itself, others imagined that the monkeys might demonstrate the natural potential of the species for rehabilitation, reproduction and recycling as research animals – an issue of interest for the pharmaceutical industry and a viewpoint supported by primate experts from Danish zoos. A consensus eventually emerged that the former research monkeys deserved rehabilitation under circumstances adapted to their natural needs.

The transfer of the Danish monkeys coincides with radical changes in the landscape of primate research in Europe over the last 20 years. Most centers have been closed down and the transition and history of such spaces are central places for investigations of the lives impacted. Our study details one specific case of disposing of nonhuman primates after the closing down of a research center. We are interested in the way understandings of good housing arrangements are co-constructed with understandings of the relationship between humans and animals. In particular, the notion of naturalness has interested us, and the way the understandings of naturalness were used by various actors to continue instrumental uses of the monkeys under new and improved conditions. Thus our paper – among other things – may be seen as a critical reflection on the views of enrichment and naturalization as unambiguous improvements of the welfare of lab animals.

In the SHH laboratory during the 1980s and 1990s, the monkeys were housed individually in narrow steel cages, fed and cleaned to the clock and treated with strongly disciplinary, sometimes violent, physical measures. Their bodies, chosen for experimentation because of their biological similarity with humans, were seen as physical objects made to serve specific scientific goals, and views on their moral value and natural needs were rarely articulated. As the transfer was being negotiated, this understanding of the monkeys as well as their treatment gradually changed, and was now increasingly seen as not only cruel and unethical but also unnatural. It was contrasted with an idyllic view of a future life in the zoo where single-cage enclosure would be substituted by a large cage with outdoor access where the monkeys' natural potential for a collective life with conspecifics and friendly relations with humans was to be developed. These two different understandings of the monkeys, their housing conditions and relationship with humans and conspecifics, were expressed in arguments presented in the debate on the future fate of the animals, in articles, interviews, observations and guidelines.

Our analysis falls into three parts. First, we characterize in more detail the two different housing arrangements in the lab and in the zoo, and the human–animal relationships taking place within them. Second, we explore the arguments for a transfer and its background. We look at the welfare provisions introduced as elements of a practice that seek to shape animal behavior in accordance with the animals' natural features in a publicly acceptable way without jeopardizing their potential value for research. Finally, we study the hard work required to turn the aged and worn-out lab monkeys into beings with a species-specific nature. Initially, however, we briefly introduce our study and our analytical framework.

Our Study

Our analysis is an ethnographic study based on interviews with a range of psychiatric and veterinarian researchers, animal technicians, hospital administrators, primate experts from Danish zoos, and representatives of animal welfare organizations and the pharmaceutical industry. Our informants were involved in the lab research during the 1980s and 1990s, in the public debate and in the rehabilitation process since 2003. We did 23 interviews in total. Some of our informants were interviewed twice, some in person, others contacted by phone or mail. We also visited the zoo to which the monkeys were transferred and observed the new housing facilities. In addition we have employed the scientific literature and other published material documenting the transfer, originating from newspapers, popular journals and scientific publications. Our empirical material furthermore includes a number of regulatory documents such as scientific guidelines and recommendations in the field of animal welfare and housing infrastructure. Our interviews all concern historic events and we see them as reflections on our informants' experiences of the past, as well as their present view of the monkeys.

Analytical Framework

In our exploration of the housing arrangements of the monkeys we consider spatial organization and human–animal relationships as co-constructed. We are particularly inspired by the animal studies tradition and the domestication studies that have emerged in recent years. Such studies highlight that, in spite of captivity, animals are not just passive objects of human manipulation but show resistance, which has effects and contributes to shaping relationships with humans (Cassidy and Mullin 2007; Candea 2010; Haraway 2007). In our case, the new housing arrangement is seen as a way of establishing a relationship between animals and humans that use this agency productively rather than repressing it.

To understand the character of such a human–animal relationship we draw on Agostin Fuentes' concept of "human niches" (Fuentes 2007). In a study of semi-wild monkeys in Bali, Agostin Fuentes has described how primates and humans co-exist in "human niches" (Fuentes 2007). He points to the importance of studying the "contexts and patterns that characterize the interface between humans and the animals they share space with" (Fuentes 2007, 124). We suggest that the lab and the zoo may be seen as such niches. Fuentes further proposes four categories for studying relations between humans and primates: *integrated*, meaning a conventional exploitative form of domestication where humans collect the payoffs from the labor of primates; *engaged*, meaning incorporation into human culture favoring certain forms of benevolent human behavior towards primates; *penalized*, meaning deleterious connections forming antagonistic relations and potential destruction of primates, and finally *idealized*, pointing to a categorization of primates as morally dignified and symbolically connected with humans (Fuentes 2007, 140). All however imply a hierarchical relationship giving humans ultimate control, only in very different ways, and with different kinds of efforts and effects. Presently we shall show how these categories are relevant in our case as they capture some basic features of the human–animal co-existence in lab and zoo.

Of special significance in our case is the role of the physical infrastructure for the way the animals and humans co-exist. The historian Reviel Netz looks upon housing of animals as a basic form of movement regulation aiming at control and ultimately value production. This regulation of human or animal movement constitutes a form of historically specific political control: "Without control over motion, value cannot be formed" (Netz 2004, xii). The value at stake in our empirical study is the medical and subsequent economic value accruing from successfully executed, socially acceptable, and scientifically productive experiments on monkey bodies. The control forms, such as the housing arrangements applied, shape the activities and practices of the humans engaging with the lab animals and influence the public view of the right way to house and handle the animals. To Netz, the structure and design of physical barriers to movement is one of the key themes in the history of the relationship between human and animal. Looking at the transfer and the rehabilitation of the Danish monkeys we consider their history as a case where different organizational and physical infrastructures are employed to regulate their bodily movement for researchers to obtain a specific kind of control. In this case the control needed to avoid public criticism, to be able to represent animals' nature in a publicly acceptable way to possibly continue experimentation with monkeys in future labs and produce valuable scientific knowledge. This implies that the transfer from lab to zoo should not be seen simply as a movement from prison to freedom, from exploitation to emancipation or from artificiality to naturalness, as it was often presented in the public debate. Instead it may be seen as a change from individual enclosure of separate bodies to collective enclosure of social groups of bodies as well as a change from one form of bodily discipline characterized by integration and penalizing human–animal relations to another, characterized by engaged and idealized human–animal relations. However, a trace of integration is retained, as the animals in the zoo are still instruments creating scientific or other value for humans.

Integration and Penalization: The Prevention of Movement and the Use of Pain

In the lab, between 1982 and 2003, the monkeys were continuously encaged and prevented from movement. The original cages in which the monkeys lived were designed as single monkey cages, 0.7 × 0.7 × 0.85 meters, some even smaller. The purpose of single monkey cages was to function as barriers (e.g. bacterially and medically) between the monkeys in the group as well as preventing escape. The cages however also prevented the monkeys' individual physical movement – a fact that had severe effects on the monkeys' bodies and behavior. Over the years this led to disability for most of the monkeys.

Prevention of movement controlled the animals 24 hours a day. Incidental aberrations from this strict control took place now and then. The cages were on wheels and in between experiments the cages could be moved either by the monkeys themselves or by a lab assistant to facilitate a minimum of physical contact between the animals such as grooming or holding hands, as a former lab assistant told us. This practice was developed among the female monkeys specifically, but their physical contact was considered undesirable although allowed implicitly from time to time. During experiments an even stronger restriction of movement was effected. Here the cage size could be reduced to an absolute minimum as one wall of the cage was moved forward to pressure the monkey body in between the walls and facilitate injections necessary for the conduct of experiments.

The monkeys at SHH were used as models of schizophrenic humans to test reactions to certain drugs; in interviews with us the scientists explained that what mattered to them was the monkeys' reaction to these drugs, which might improve the lives of psychiatric patients. Fulfilling the monkeys' basic biological need for food and shelter was a practical issue to make experimentation possible. In conversations about the welfare of the monkeys, our informants explained that the life of the monkeys in the lab had certain advantages over that of the jungle. Hunger and violent attacks from conspecifics were prevented and the physical health of the monkeys was secured. The shelter, food, and entertainment of the animals were not considered scientifically important to, nor integrated into, the research process or the scientific results. The monkeys were configured as model organisms, as bare, generic life, with basic biological similarities to humans. These similarities (e.g. their naked hairless faces) were the reason for the choice of these specific monkeys to model human facial contractions. The housing regime had adverse effects such as stereotypical behavior, pain or anger, but this was considered a separate practical problem to be solved by the technical staff (Friese 2013). Thus the housing arrangement chosen was seen as distinctly separate from the animal body and its instrumental value.

As members of the experimental animal population the monkeys in SHH were fully integrated in human society, but separated from humans and conspecifics by steel bars. This housing arrangement may be described as penalization, as humans produced value from their participation in experiments for more than 20 years and continuously inflicted suffering upon the animals, who lived in painful isolation, subject to experimentation at any time. The scientific publications resulting from the monkey studies illustrate the way the experiments were carried out, the resulting distressed behavior of the monkeys and the detached approach of the researchers:

> [The medication] resulted in an unexpectedly violent reaction, not seen before with any other DA agonist. The animals ran about the cage, at times flinging themselves against the cage. This reaction reached a peak at 30 min after injection. In the aftermath, one animal collapsed into a coma and died hours later, the most likely cause of death being cranial trauma. Due to this event, further examination of the effects of SKF 83822 [the technical name of the drug] was limited to the use of .35 mg/kg, a dose producing significant but moderate effects, and using only potential antagonists in further exploring this drug's effects.
>
> (Peacock et al. 1990)

> All of the monkeys were jittery, reacting to minimal stimuli such as finger snapping or the noise caused by the other animals' activity. Two of the monkeys were extremely anxious, pulling desperately at the cage during the entire observation period. … One of the monkeys reacted by biting the cage.
> (Peacock and Gerlach 2001)

In his exploration of the role of barbed wire in American culture, Netz argues that the control of movement originates in "the power of steel over flesh." Compared with barbed wire however, stainless steel, the material of which the cages was built, represents a refined form of movement prevention. In the lab the restriction of movement and threat of pain, however, were seen as necessary to limit the active resistance of the animal and make the animals obey the research needs of the scientists. In interviews with us, the researchers admitted that force was continuously used to access the potentially disobedient bodies of the monkeys. In addition to intensive control of the animals' movement in the narrow cages, scientists and lab technicians used additional force when performing injections and other forms of intervention. In interviews we were told that for the males, life in the narrow cages led to dramatic reactions as they were caught between an impulse to move and the limiting effects of the preventive measure. Discipline was closely linked to the specific way that space was organized. The female monkeys reacted differently, we were told. They seemed to succumb to their destiny, often put forward their arms for injection in resignation when experimentation was to begin, but also tinkered with the narrow possibilities they were allowed. They sometimes managed to move their cages enough to reach hold of each other's hands for comfort. "They liked to hold hands," explained the (female) lab assistant. These different responses to caging and isolation illustrated the spectrum of possibilities and choices available under very restricted conditions. It should be mentioned that the females were considered of less interest to the rehabilitation plans than the males. They were unable to reproduce and were euthanized just before the transfer, as they were judged to be beyond repair.

In this experimental set-up, the ideal scientist would behave distantly and matter-of-factly towards the animals. Some of the psychiatric researchers admitted the difficulties of doing this, especially in the face of animal suffering, but all explained the necessity of keeping personal sentiments and emotions under control. Trained as lab physicians in the positivist tradition, their ideals of emotional and physical distance between researcher and research animal resulted in a housing arrangement characterized by anxiety, unrest and noise on the part of the animals, and rational calculation repressing undesirable emotions on the part of the humans.

Naturalization of Animal Welfare

In the early 1990s, intensive international work took place to revise the inadequate European guidelines drawn up in the 1986 EU directive (EU 1986; Jennings 2010). Eventually, new guidelines were decided upon requiring that "animals should be kept in environments that respect their needs, including their needs to show certain behaviours." (European Commission 2002, 6). A widespread criticism of narrow, individual cages used everywhere in the medical industry had been accompanied by ethological studies demonstrating that the acknowledgement of animal needs could easily be united with efficient lab procedures. It was argued that conventional housing arrangements stressed animals and created bad publicity, whereas enriched housing arrangements adapted to natural behavior would create a happy public, happier animals and better research results (Mikkelsen et al. 2010). Already by 1970 it was documented that the performance of stressed animals in pain was not to be trusted and might skew scientific results. In his critical volume on animal experimentation from 1984, published when the first experiments at SHH had begun, Andrew Rowan cites one physiologist experienced in animal resuscitation, M. Hillmann: "There is hardly a single organ or biochemical system in the body which is unaffected by stress. …The changes in some of the parameters [of a specific experiment] may be up to 300% of the normal values at rest" (Rowan 1984, 87). Since then an increasing acknowledgement that the quality of housing and husbandry has had a major impact on laboratory animal health, and welfare has emerged (Ottesen 2004). An expanding body of scientific literature demonstrates that environments that do not meet animals' physical, behavioral and/or social needs may result in changes in physiology and to abnormal behavior (e.g. stereotypies), which not only compromise animal welfare but also influence the validity and reproducibility of the scientific data obtained (Jacobsen 2009; Katsnelson 2014).

At SHH, however, the researchers continued their conventional way of experimentation until the late 1990s, with the active support of the official Danish authorities of the Animal Welfare Agency. By then the predicament of the monkeys had become publicly known and the time had run out for SHH. Two highly influential actors, the Danish pharmaceutical company Novo Nordisk and the Danish Animal Welfare Society *Dyrenes Beskyttelse* (DB), had begun to ventilate critical views of the conventional single-cage system. Out of a long-lasting dissatisfaction with the poor housing conditions offered to research animals such as rabbits, dogs and guinea-pigs, these two organizations proposed to investigate the basic needs (Ottesen et al. 2007, 397) of research animals, and the development of prototypes for housing to meet such needs. Together they campaigned for better animal welfare in the research lab with the motto "Animal welfare – a question of respect." The main viewpoints of the campaign were that improved environmental conditions such as housing and husbandry were vital elements in the welfare and usefulness of lab animals and that automatic killing at the end of an experiment was unnecessary.

Financed by a generous budget from Novo Nordisk, a number of workshops were organized, inviting internationally recognized experts "to establish the most important needs of the animals in their natural habitat," a new baseline for naturalness, so to speak (Ottesen 2004, 1). The approach was christened the "near-to-nature" system by the British ethologist, Marcus Stauffacher, who was part of the group that Novo Nordisk and DB brought together. According to him, "ethological examination of group breeding has shown that this housing system is both more suitable to the requirements of [the animals] themselves and more in the interest of breeders than the present commercial cage system" (Stauffacher 1992, 105). The nature spoken of here was constructed on the basis of ethological studies of the behavior of the animals under various circumstances and with a range of options for the animals to choose from. The aim was to allow the animals to live together "without friction" (Stauffacher 1992, 112). The workshop concluded that social contacts with conspecifics (group housing), freedom of movement, enrichment and consideration of individual species' needs were important elements. New housing prototypes were proposed for various species, and recommended as new standards to be tested scientifically (Stauffacher, 1992). This collaborative work, which lasted from 1999 to about 2003, forms an important backdrop to the decision to transfer the SHH monkeys to a zoo. Novo Nordisk decided to donate 200,000 DKK to the rehabilitation project, and DB organized a campaign collecting two million DKK to finance the design and building of a modern collective cage (Mikkelsen et al. 2007; Ottesen et al. 2007).

In an interview with us, the former head of Novo Nordisk's ethical section described how one of the results of Novo Nordisk's interest in new housing systems was translated into the building of new stables at Novo Nordisk for dogs and rodents. Naturalistic housing arrangements were expected to create a whole new relationship between animal and researcher. She told us that the new trend implied that animals could be trained to collaborate with scientists. As an interesting parallel to the subjected female monkeys in the old single-cage regime who were resigned to collaborate under threat of violence, the interviewee explained that animals living under "near-to-nature" conditions could be trained to voluntarily stretch out their hands when the scientists needed a blood sample without any use of violence. This image of friendly collaboration between humans and animals forms a dramatic contrast to the anxiety-ridden scenery described by several psychiatrists when remembering their experiments in the 1980s at SHH. It was part of the emerging international trend in animal welfare research and practice towards creating "naturalistic environments" for lab animals (Ross et al. 2009, 42), characterized by attention to species-specific needs, environment and behavior. As a response to the longstanding demand for refinement of animal experimentation (one of the 3 Rs), representatives from both the Danish pharmaceutical industry and Danish animal welfare organizations were active in international reform work to create and implement new guidelines to "naturalize" housing arrangements for lab animals.

Engagement and Idealization: "Naturalistic" Housing and Enabling Movement

With the decision to move the monkeys to a collective cage with access to a much larger area, including outdoor access, the human actors behind the transfer enter into a different relationship with the monkeys. The chosen zoo was situated in a rural part of Denmark. The new cage was designed by an architect in collaboration with the zoo staff and animal welfare organizations, and took animal needs and public views into consideration from the beginning. The ambition was to "take our starting point in nature," as the head of the zoo told us. The cage was designed according to the tree-top principle, a carefully thought out arrangement, where monkeys could gradually get used to socializing with each other – in pairs and foursomes, but also allowing the weaker monkeys to escape to more distant quarters, difficult to access by dominant monkeys. They had access to an outdoor island where they could roam freely – protected from unwelcome human (visitors') interference. This new housing arrangement reflects a new understanding of animal nature: the unit of interest was no longer the isolated physical animal organism, but the animal plus conspecifics plus environment. Housing was no longer solely a means of assisting human control of the animal but had become part of the animals' redefined nature. The monkeys were now seen as members of a species with specific needs to be cared for. The effects of their original confinement in narrow metal cages were sought to be reversed, a change that required abandonment of the habitual physical separation in individual cages and a gradual adaptation to living in a collective cage and being in physical proximity to others. The monkeys were served natural foods to accommodate their natural tastes and needs just as the food was artificially made more difficult to access – to more closely resemble natural conditions.

After the transfer the monkeys undoubtedly were exposed to less brutality – even though open collective cages should not be mistaken for freedom. This housing system implies human–animal relations characterized by a benevolent and friendly incorporation into human culture, and a view of the animals as subjects of respect and dignity and deserving a natural life. A more open collective housing arrangement allowed for a different kind of primate agency where the monkeys recruited humans such as researchers and conservationists into their own social dynamics – only now in a more friendly and non-violent way, but still compatible with attempts to help satisfy human needs, illustrating the persistent trace of integration in the new lives of the monkeys. The enablement of movement was a vital element in the emerging understanding of what "natural" meant, and signified liberation from imprisonment, restriction and compulsion. Also, "naturalness" implied that death was put off indefinitely – although within the domain of human decision-making. Death from old age was considered preferable. In other words, the naturalized arrangement mimicked the ideal human life. Drawing upon Fuentes, we call this relationship "engaged idealism." It is engaged because the primates were now incorporated into human culture in a way that was explicitly motivated by a moral impulse to provide better conditions for the animals. And it is idealized since the view of these animals builds upon an ideal of primates as representatives of nature deserving a natural life. The monkeys retained their biological similarity with humans at the same time as they left their former antagonistic relationship with them and became partners in a new moral project.

The transfer may be seen as a qualitative shift in the history of the SHH monkeys, as well as indicative of a shift in the relationship between animals and humans, including the public. This shift was largely made necessary by the revelation of the conditions of the monkeys to the public by exposure of their predicament in the lab – as illustrated by the scientific publications quoted above. Today such detailed description of the suffering of a lab animal is unimaginable and in hindsight it seems almost naïve. Thus, in the lab, where a conventional research design was adhered to, co-existence took a specific form, a combination of integration and penalization. Situated within an important scientific institution producing vital value for humans, the social and physical distance between animal and human was pronounced, and the infrastructure allowed for strict disciplinary control of the monkeys, leading to violent resistance or in a few cases complete subjugation. Animal welfare was narrowly interpreted in terms of basic material needs such as food, water and basic care. In the zoo, co-existence was characterized by engagement and idealization. Simultaneously humans engaged in benevolent relations with primates, seeking to accommodate their needs and incorporating them into human culture in various ways (e.g. via the media, ethical guidelines and public engagement).

Back to Nature

In Western culture, "nature" and "naturalness" are counter-concepts to control and exploitation. Therefore references to naturalness certainly had the effect of making further instrumentalization of the SHH monkeys less visible. In this light, animal welfare provisions such as enriched housing may be seen as a governmental practice not exclusively done for the sake of the animals, but in ways compatible with scientific and commercial needs of the industry. The enlargement of the cages to spacious and "natural" surroundings for social groups of animals is one such enrichment.

"Animals can at least go as far as the fence," says Netz in his discussion of the qualitative change of the forms of control he claims took place in the mid-twentieth century – and the fact is that increasing numbers of research animals in the West are today fenced rather than caged (Netz 2004, 238). In the Danish case, the actors involved in the transfer pictured the zoo as a space where monkey nature might unfold itself. They presented the transfer as a humanitarian way of saving the lives of the monkeys, an act of responsible goodness, which contrasts with the allegedly terrible conditions in the lab. In real life, this ideal natural life of the monkeys turned out to be more difficult to transform than expected. One veterinarian researcher and the primate expert from the zoo described the adaptation to new housing conditions as an ordeal, a process that had to be performed carefully and first of all slowly, to reduce the amount of stress, pain and anxiety associated with the abandonment of the habitual life as research monkeys. One primate expert explained how an assistant from the zoo arrived a couple of days before the transfer and spent time with the monkeys at Skt. Hans and helped make them used to the transportation boxes used for the transfer which took place in a preheated car. Nevertheless the new circumstances created frustration and conflict among the monkeys. They had to leave the familiar single cages, learn to eat "proper" monkey food (in the beginning the monkeys were afraid to eat live food such as worms), getting to know each other physically (in the beginning there were fights) and adapt to a collective cage where the social hierarchy could be played out in a novel – and to some, more threatening – manner. The head of the zoo explained:

> They acted like children, were totally unruly, they had been sitting for fourteen [some for 20] years closely together without ever being able to act out. Social monkeys have a behavioral pattern which implies communication, and when you entered their stables they got so excited that it was next to intolerable to be there without hearing protection.

He argued for a prolonged adjustment period to teach the monkeys to behave naturally: "They had been induced permanent psychoses, they were chronically damaged from being bombed back to the stone ages … their ability to move was really, really bad, their muscles had been totally tense and they just functioned very badly." To soothe the stressed and often frightened monkeys the staff drugged the monkeys during the first weeks: "We had to use some chemistry [anxiety-reducing drugs] … many of these monkeys were scared to death." For no obvious reason one monkey died just before the transfer. Another had to be put down just after the transfer after another monkey had pulled off his arm. When they arrived to the new cage, and were let out of the transport boxes most of the monkeys were either physically unable or afraid to climb up the trees designed for their "natural" behavior, and several slept on the floor during the first nights. Only after 14 months in transit where the monkeys were gradually adapted to their new conditions could they be let out into the new collective cage. "Their worldview changed drastically … but very quickly we saw that they adapted to the tasks we gave them, and their stereotypes began to disappear. First during limited intervals, then – for some of them – completely." Not all monkeys were able to adapt to the new life. Of the original twelve monkeys who were transferred only five were alive when we visited the zoo on February 16th 2012. Today only one remains. Natural behavior obviously had to be "re-learned" and the care taken to attend to and re-establish monkey nature, turned out to be hard, protracted work.

The Novo Nordisk Study

The transfer had the prime ambition to enable movement without compromising control, but the animals were still enclosed, only in a larger more comfortable cage. For some of the actors in the alliance this large collective cage was seen as a prototype for future housing arrangements in an enriched experimental setting where man and animal would collaborate in a friendly manner, using the natural capacities of the animals in the interests of man. In Novo Nordisk, the company's ethicists theorized that the monkeys had been exposed to a suboptimal life quality and housing conditions and with a group of monkeys with stereotypes at hand, the opportunity presented itself to do specific experiments to find out what it would take to possibly eliminate this behavior. This view supplemented the original decision to support the housing of the monkeys in a collective cage, a decision that would make it possible to experiment with shifting housing regimes – between "natural " and "experimental" conditions and back again.

In 2008, Novo Nordisk financed a small study to find out whether the new housing arrangement provided an enriched environment that would lead to changes in behavior and physiology indicating improved welfare (Jacobsen 2009). The head of the ethics section contracted a veterinary student to perform an observation study of the monkeys' behavior during the initial phases of their new life. The aim was to find out if knowledge of the new housing arrangement might feed positively into development of housing plans for animals used in overseas contract labs, where large Danish pharmaceutical companies like Novo Nordisk and Lundbeck were now conducting their primate research. However, these overseas partners would have to comply with a predefined set of ethical principles. The argument was that in spite of the increased costs of better animal welfare (larger cages, enriched housing, etc.) the benefits such as public goodwill, better occupational health of lab technicians and more reliable scientific results would counterbalance the costs. Even though primate research is severely criticized by animal welfarists, it is still considered vital for specific kinds of research such as autoimmune and neurological diseases. For the international big pharmaceutical industry, including Novo Nordisk, the rehabilitation of the SHH monkeys may have been a minor event, but it was still a chance to investigate whether enriched housing conditions would further more "natural" behavior, compatible with advocating and supporting necessary primate research in accordance with monkey nature and needs. This was not particularly well documented when it came to animals in need of rehabilitation but was highly interesting to the actors who had an interest in recycling lab animals rather than killing them.

This small behavioral study performed on the SHH monkeys also provided information on what would count as success in rehabilitation attempts. The young veterinarian who performed the Novo Nordisk study was very skeptical of the new housing system: "I don't think they realized how damaged they were. Single caging is the worst one can do to a monkey. These males have been sitting in cages facing each other and puffed themselves up to each other for 8 to10 years and then they are suddenly placed in the same cage. That is a situation they can't cope with." Still, her study concluded that even the older monkeys had "sufficient cognitive flexibility to express behavioral profiles that resemble those of their wild conspecifics."

For the medical industry as well as for the zoo, the result can hardly be called an unqualified success (Alstrup 2009). In spite of the establishment of the new housing system, the original plans of turning the monkeys into a valuable genetic resource fell to the ground. None of them were transferred to other zoos, and the planned breeding program was eventually abandoned. The initial hopes of full recovery were not met, stereotypical behavior persisted in most monkeys and several monkeys died. Yet the ambition among some actors to use the new set-up as a knowledge-producing arrangement serving human purposes was intact, and the view persisted that the transfer was the ethically right thing to do. The support of the rehabilitation project was certainly beneficial for Novo Nordisk as well as for DB, regardless of the possible future uses of these specific monkeys. The projected naturalness of the monkeys helped make these organizations appear as socially and ethically responsible actors on the public scene and the animals were turned into their allies in this project.

Conclusions

In her book on the use of animal and mechanical parts in transplantation practice, Lesley Sharp uses the concept of naturalization, as she claims that lab animals are "naturalized" in the sense that they are no longer wild, but have changed their status from wild to "civilized." We have taken her use of this concept further, and claim that the lab animals from SHH were sought to be re-naturalized as they were moved from lab to zoo; this movement did not mean that they were no longer scientific objects, nor that they should be seen as wild animals. On the contrary, the plan was to allow them to gain a second nature, so to speak, neither wild animals nor conventional lab animals, but rather semi-natural. As naturalization requires deliberate efforts to alter the "nature" of the wild animal (originally the monkeys underwent a strict purification regime to become accessible for scientific experimentation), the re-naturalization process required similarly strict procedures to re-make the lab animals fit their new status as potential experimental zoo animals. These transformations included hard work, for animals as well as for humans, and the value of the transferred animal rested strongly on its ability to adapt to the new conditions and approximate its new "re-nature." The transfer itself was presented as a moral project, and the zoo may be seen as a moral laboratory experimenting on the right methods of re-naturalizing worn-out lab monkeys (Mattingly 2014). This experiment rested upon an idealization of the monkeys as deserving relatives of humans. In this process humans understood themselves as engaging in a moral project, doing good to animals and establishing idyllic relations with animals, even as managed and fully controlled objects of experimentation.

In Western culture, the notion of nature most often refers to the wild, and is seen in contrast to culture, civilization, as shaped by humans (Ingold 2000). But as humans increasingly transform nature into culture, the ambition to naturalize contains a deep-rooted contradiction (Sharp 2014). No longer is nature the opposite of human culture but is deeply culturalized. The opposition has been dissolved, and appeals to nature and naturalism seem today to be associated with a better way of being human. In our case, we propose that "nature" was put to work once more to serve humanity – now by rhetorical means. With Lesley Sharp's expression we may speak of "a politics of the idea of nature" (Sharp 2014). Paradoxically this politics made it possible to defend nature and exploit it at the same time. The scientized concept of "near-to-nature" as it emerged from the animal welfare researchers characterized a reformed housing regime shaped by and serving current human interests. As we have shown above, the new understanding of monkey nature and its naturalistic environment requires intensive culturalization, which may be described as far removed from the wild, i.e. re-naturalized or "cultured-up" (Franklin 2013, 4; see also Sharp 2014). In this case *naturalness* reflected human notions of nature that corresponded with what made humans feel good. Thereby our study demonstrates how hard it is to establish *natural* conditions. The animals were not easily adaptable to new conditions after many years of captivity, regardless of the naturalness of such conditions. Animal nature is not intrinsic but should be seen as a social construction, a result of training, domestication, and inherently shaped by specific human–animal relationships.

Acknowledgments

Our warmest thanks go to our informants, who contributed their experiences and reflections; to the participants of the Oslo Workshop on Animal Housing in 2013 for their comments on an earlier version of the paper; and to Carrie Friese and Rob Kirk for invaluable support. Last but not least we thank our reviewers for their critical scrutiny and encouraging reviewer comments.

References

Alstrup, A.K.O. (2009) "Var flytningen af kapucineraberne fra Sankt Hans Hospital en succes?," *Dansk Veterinærtidsskrift,* 92: 16–19.

Candea, M. (2010) "'I Fell in Love with Carlos the Meerkat': Engagement and Detachment in Human-Animal Relations," *American Ethnologist*, 37: 241–258.

Cassidy, R. and Mullin, M. (2007) *Where the Wild Things Are Now: Domestication Reconsidered*. Oxford: Berg.

European Commission (2002) *The Welfare of Non-human Primates Used in Research. Report of the Scientific Committee on Animal Health and Animal Welfare*. Brussels: European Commission.

EU (1986) EU Directive: Council directive 86/609/EEC.

Franklin, S.B. (2013) *Biological Relatives: IVF, Stem Cells, and the Future of Kinship*. Durham, NC: Duke University Press.

Friese, C. (2013) "Realizing Potential in Translational Medicine: The Uncanny Emergence of Care as Science," *Current Anthropology,* 54: S129–S138.

Fuentes, A. (2007) "Monkey and Human Interconnections: The Wild, the Captive and the In-Between," in R. Cassidy and M. Mullin (eds), *Where the Wild Things Are Now: Domestication Reconsidered*. Oxford: Berg, 123–145.

Haraway, D.J. (2007) *When Species Meet*. Minneapolis: University of Minnesota Press.

Ingold, T. (2000) *The Perception of the Environment: Essays on Livelihood, Dwelling and Skill*. London: Routledge.

Jacobsen, K.R. (2009) *Miljøberigelse af kapuchineraber*. Københavns Universitet. Det Biovidenskabelige Fakultet for Fødevarer. Veterinærmedicin og Naturresourcer. Institut for Veterinær Sygdomsbiologi.

Jennings, M. (2010) "Special Protection for Primates – the Need for Faster Progress," *AATEX,* 27: 185–189.

Katsnelson, A. (2014) "Male Researchers Stress out Rodents: Rats and Mice Show Increased Stress Levels When Handled by Men Rather than Women, Potentially Skewing Study Results," in *Nature News and Comments*, 28 April 2014. [Accessed May 20, 2015].

Koch, L. and Svendsen, M.N. (2015) "Negotiating Moral Value: A Story of Danish Research Monkeys and Their Humans", *Science, Technology & Human Values*, 40 (3): 368–388.

Mattingly, C. (2014) *Moral Laboratories: Family Peril and the Struggle for a Good Life*. Oakland: University of California Press.

Mikkelsen, L.F., Hansen, H.N., Holst, L. and Ottesen, J.L. (2007) "Setting Global Standards for Animal Welfare Monitoring of External Contractors," *AATEXX,* 14: 731–733.

Mikkelsen, L.F., Sørensen, D.B., Krohn, T., Lauritzen, B., Dragsted, N., Hansen, A.K., et al. (2010) "Clinical Pathology and Cardiovascular Parameters Are Not Influenced by Housing Rats under Increased Environmental Complexity," *Animal Welfare,* 19: 449–460.

Netz, R. (2004) *Barbed Wire: An Ecology of Modernity*. Middletown, CT: Wesleyan University Press.

Ottesen, J.L. (2004) "New Housing Standards for Rats and Mice Developed with Focus on the Needs for the Animals" in *The Development of Science-based Guidelines for Laboratory Animal Care: Proceedings of the November 2003 International Workshop*. NCBI Bookshelf.

Ottesen, J.L., Mikkelsen, L.F., Bertelsen, T., Krohn, T., Velschow, S., Hau, J., et al. (2007) "Collaboration between Academia and Industry with Focus on Improvement of the Welfare of Both Animals and Humans in Laboratory Animal Facilities," *AATEXX,* 14: 69–71.

Peacock, L. and Gerlach, J. (2001) "Aberrant Behavioral Effects of a Dopamine D1 Receptor Antagonist and Agonist in Monkeys: Evidence of Uncharted Dopamine D1 Receptor Actions," *Biological Psychiatry,* 50: 501–509.

Peacock, L., Lublin, H., and Gerlach, J. (1990) "The Effects of Dopamine D1 and D2 Receptor Agonists and Antagonists in Monkeys Withdrawn from Long-Term Neuroleptic Treatment," *European Journal of Pharmacology,* 186: 49–59.

Ross, S.R., Shapiro, S.J., Hau, J., and Lukas, K.E. (2009) "Space Use as an Indicator of Enclosure Appropriateness: A Novel Measure of Captive Animal Welfare," *Applied Animal Behaviour Science,* 121: 42–50.

Rowan, A.N. (1984) *Of Mice, Models, and Men: Critical Evaluation of Animal Research*. Albany: State University of New York Press.

Sharp, L.A. (2014) *The Transplant Imaginary: Mechanical Hearts, Animal Parts, and Moral Thinking in Highly Experimental Science*. Berkeley: University of California Press.

Stauffacher, M. (1992) "Group Housing and Enrichment Cages for Breeding, Fattening and Laboratory Rabbits," *Animal Welfare,* 1: 105–125.

6 Housing Eiders – Making Heritage

The Changing Context of the Human–Eider Relationship in The Vega Archipelago, Norway

Knut Fageraas

Eider ducks (*the common eider*) are found to be quite tame. For centuries they have been nesting in simple human-made shelters along most of the North Atlantic Coast of Norway. This large sea duck, quite often referred to as a farm animal, was most commonly found to populate the numerous small islands and islets of the vast Vega Archipelago at Helgeland in Nordland County. The eiders seek to be close to humans living within the island realm, making use of the nesting houses (*e-hus/e-bane*) to have better protection against their predators. The human population of smallholders built the nesting houses for easy access to the eggs and down. The eider down is considered to be the finest of all down, and was an important source of income to the smallholders living within the archipelago. It is collected from the nests after the hatching period, cleaned and used as filling for duvets and pillows. From the 1500s onwards these eider down products were exported and sold as luxury products to noble and royal houses around Europe.

The smallholders living in the Vega Archipelago, the so-called fisher-farmers, based their livelihood on fishing, small-scale farming, and collecting eggs and down from wild birds. To them the eider down production was an important economic activity that, for some, could equal the value of the seasonal fishery in Lofoten, the world's largest cod fishery. Unlike fishing, the eider down production, along with most other farm duties and household tasks, was left to the women. The hard work of bringing up seaweed from the shore to prepare as nests was entrusted to men. The crafting of nesting houses was also often work done by the men. Nonetheless, the main occupation of the men was fishing, whether in the near-coastal area or the far-off fisheries of which the Lofoten fishery was only one of many. The eider was a privileged bird species compared with most animals due to its significance to the fisher-farmer's livelihood. It was considered to be a sacred bird that was not allowed to be hunted, and which people took every possible precaution to protect and care for so that as many eider ducks as possible would prefer to hatch in the human-made sheltered nests.

Figure 6.1 Eider drakes and an eider duck in the inlet of Lånan. On shore, in the background, is one of the numerous nesting houses (e-bane) on the site

Source: Photograph by W. Dramstad, Lånan, 2009/© Skog og landskap.

The tradition of housing eider ducks and producing eider down still continues on the outermost islands in the Vega Archipelago. However, literally all the traditional activity was abandoned during the post-war period due to depopulation of the small island settlements. Most of the eiders in the archipelago disappeared in the same period. Still, the last two decades have seen a revival of the tradition, mainly on islands where the largest eider colonies in the past had stayed and been cared for by humans. Interest has grown, particularly after *The Vega Archipelago* was listed as a UNESCO World Heritage Site in 2004 and the World Heritage Committee decision emphasized mainly the tradition of "eider down harvesting" and "the down process" (WHC 2004). Consequently, as the tradition depends on the presence of eiders, these birds have now regained a privileged position. The revival of the tradition of housing eiders in recent years is believed to have increased the eider population in the archipelago compared with a historically low number in the 1990s.

This paper presents the human–bird relationship produced within the traditional practice of housing eider ducks in the Vega Archipelago. It deals with a way of humans living with wild birds about which there exist very few written accounts. Most accounts of the human–eider relationship were written between 1996, when Vega was proposed as a potential World Heritage site, and 2004, when the status was awarded. The writing comprises both local historical literature and official documents, such as the nomination file proposing the area for the World Heritage List. I regard all such writings as the result of the heritage discourses that occurred prior to as well as in the context of the area's World Heritage nomination process. In these writings, the human–eider relationship has been described in different terms, ranging from animal husbandry to interspecies interplay. The eiders have been termed as both wild and tame; and characterized as non-domesticated species as well as regarded as included in humans' domestication practices. But what are the locals' actual experiences of the human–eider relationship? Are they different from the way they are being reproduced in the heritage discourse? What do people actually know about how the eiders relate to humans? Are the human-centered ways of thought capable of explaining the birds' behavior and sentience? These are the research questions I have been occupied with.

I start with a theoretically based reflection on human and nonhuman animal relations. This I use as a conceptual frame in the successive sections to present the tradition of building eider duck nesting houses, its character as a domestication practice, and how people adapt to the presence of these birds in their environment. I then continue by reflecting about how the human–eider relationship is understood by the local enthusiast as well as the heritage experts, in order to reflect upon the changing societal context of the humans' traditions in relation to the nesting eider ducks. My analysis is, however, primarily based on the expressions of those with actual experiences with the eider-related traditions. As sources of local knowledge they have been influencing the heritage discourse – and are themselves influenced by it, not least, as motivation for reviving the tradition relating to the eiders. I am inspired by the scholarly interest within Nordic ethnology in animal living as it takes place within human environments, which has grown in accordance with the broader evolving field of interest, often referred to as the *animal turn*, in the humanities and social sciences during recent decades. Moreover, I draw upon the field of critical heritage studies, in which theories of discourses are central in explaining how values and understandings of heritage are construed. Simply put, and based on the discourse theory of Ernesto Laclau and Chantal Mouffe (2001), discourses are specific ways people talk about, understand, and interpret the surrounding world (Jørgensen and Philips 2002, 1). At the same time, discourses create and change the common understandings of the subject matter under consideration (ibid.). There can be several different, and competing, understandings/discourses within the same discursive field.

The paper is based both on ethnographic fieldwork and on written sources such as local historical literature and policy documents from the national and international heritage authorities. The primary source material is semi-structured interviews with forty people at Vega, undertaken in 2008 and 2009, and follow-up talks in subsequent years. Some of the interviewees have key roles in the management of the World Heritage Area, while others are members of NGOs supporting the continuing of the eider-related traditions. Ten of the interviewees are property owners on the outer islands or are related to these sites through family members or their ancestors. Parts of the year they stay at the outer islands in order to care for the nesting eider ducks. All of these ten interviewees have grown up on small islands in the archipelago with actual experiences of the way of life where eiders had great significance. They have also been central to the revival of the eider-related traditions. Three of them have upheld the traditions throughout most of the post-war period, but only one has continually kept on regardless of others who occasionally stopped. The 1980s and 1990s were particularly critical periods with rather few and mostly elderly practitioners and a small eider population. At that time, there were only 3,000 eider ducks breeding in the archipelago, which was about one-tenth of the number in the late 1800s. Most of these birds nested in the open landscape, with a small number in the relatively few human-made shelters.

Perspectives on Human Relations to Other Animals

The practice of housing animals tells a lot about the relationship between humans and nonhuman animals. It can be conceptualized through the long tradition of domestication. Cassidy and Mullin (2007) explain domestication through its 10,000 year history (which is as long as people have lived at Vega), which has produced a number of intended and unintended consequences that continue to resonate today (cited in Wilkie 2010, 17). It is essentially about the domination by people and their culture over other animals and nature, signaling a clear distinction between the humans and nonhuman animals. According to the sociologist Rhoda Wilkie, however, practices of domestication can appear in radically different ways. They can appear in the form of a "social contract," typically characterizing the productive relationship between traditional farmers and their animals; or the total control and dependence that rule methods of intensified animal production of the post-war period (ibid.). The relationship between humans and eiders in the Vega Archipelago is, in the heritage discourse, represented in a manner similar to people's animal keeping in traditional societies.

It can, however, be expected that the eider ducks relate differently to other animal's presence than humans do. Jacob von Uexküll, who has been called a "preeminent founder of modern ethology," forwarded a novel theory in the study of animals' perception of their surroundings (Sebeok 1988, 66). It is based on a phenomenological ontology relevant for all living species – including humans. Uexküll (2001 [1936]; 2010 [1940]) used the term *Umwelt* to conceptualize how the surroundings and the animal constitutes a relational holistic unitary. Objects encountered through the animal's "stroll" through life have significance by acquiring, or performing, a functional "quality," which differs with various species (and individuals) (ibid.). This theory has inspired a growing number of later researchers who have interest in animal life, like the anthropologist Tim Ingold:

> Central to his [von Uexküll] approach was the idea that the animal, far from fitting into a given corner of the world (a niche), actually fits the world to itself, by ascribing functional meanings to the objects it encounters, thereby,

> integrating them into coherent system of its own. These meanings, he insisted, were not given in the objects themselves, but were acquired by those objects by virtue of their having entered into a relationship with an animal subject.
>
> (Ingold 1988, 13)

According to this ontology, the activity of humans is fitted into the nonhuman animals' *Umwelt.* Subsequently, to the eiders, the human-made eider duck nesting houses, as well as people's adaptive actions in the birds hatching period, perform a functional *quality* as means of protection.

Ingold exemplifies Uexküll's concept of *Umwelt* by contrasting it with the ecological psychologist James Gibson's concept of "affordance." To Gibson, objects have invariant properties independent of how they are used by different animal species. This would mean, Ingold infers, that animals in a shared environment potentially also would share perceptions of what objects afford, and, also, that this conception implies a form of social relation among different species. For example, in my case, such a conception would imply that humans and animals would share the perception of materials being feasible for making protective constructions; furthermore, that their relations were structured by a kind of social contract. The human–eider relationship can be interpreted in such a way, whereas the eider ducks, in return for protection, provide the humans with eggs and down.

For Uexküll, however, any animal (humans as well) is enclosed in its own subjective universe; their "phenomenological world" or "self-world," which is only accessible to themselves (Uexküll 2010 [1940], 5; cf. Ingold 1988). If so, and in line with the historian of science, Lorraine Daston (2005), for humans, any attempt to understand other animals species' experiences ought to be marked by anthropomorphism. Such attempts often lead to perceptions that humans and animals share thoughts and feelings (Daston and Mitman 2005). Hence, the perception of interplay between people and eiders in the Vega Archipelago, connoting shared interspecies sociability, feelings, or a social contract between them, is based on anthropomorphism. However, the relationship with humans can, in fact, rather be conceived to be unnatural to the eiders, as for all other nonhuman animals.

The ethnologist Liv Emma Thorsen (2006), referring to the historian Nigel Rothfels, defines nonhuman animal presence in any context of the human environment as concerning their "unnatural history." She sees the domesticated animals as the perfect example of what is unnatural to the animals:

> The core of the animals' unnatural history is its aspect of being relational and transgressing. When animals are literally entered into cultural history, it is as a relationship between the animal and man, a relation which holds a reminder of what it means to be a human and that animal lives are being lived in parallel to our own.
>
> (Thorsen 2006, 25 [my translation])

The conception of the nonhuman animal Other has led Thorsen (ibid., 29ff) to explore occasions of animals meeting with humans by analyzing their presence within human environments. It is a battle about territories, she explains, inspired by the cultural historian Troels-Lund's (1968 [1908–10]) epic work about the daily life of humans and animals in the Nordic countries. Troels-Lund describes the animals' behavior from their point of view. To him they were the Others, battling human expansion into (previously) the animal territories as civilization has developed. I use this kind of territorial thinking, the conception of nonhuman animal otherness, and the perspective of their unnatural history, to reflect broadly on the practice of housing eiders and how it is valued and understood in the context of its World Heritage significance. This way of thought is, as I see it, not too far from the Uexküllian ontology. I use the theoretical basis of this thinking to explore the complexity of relations between the human and the eiders in the Vega Archipelago.

The Tradition of Building Eider Nesting Houses

It is believed that the practice of building nesting houses and producing eider down originates from the southern region of North Norway – quite likely from the coastal region in Nordland. This has been the core area through the centuries (Suul 2012, 45/62). It was probably most customary on the small outer islands

Figure 6.2 Some of the eider ducks allow the island dwellers to touch them when they are sitting on the eggs

Source: Photograph by K. Daugstad, Lånan, 2009.

and shallow waters in the Vega Archipelago, which provided ideal living conditions for the eiders (ibid., 73). In 1900, Nordland produced a ton of cleaned eider down, two-thirds of it in Helgeland, where the Vega Islands were the most productive area with 118 kg (Næss 2013, 151). The tradition was brought to Iceland and Greenland as early as the 800s. At present Iceland is by far the biggest manufacturer of eider down with 2.5 metric tons of mechanically cleaned down product each year. In Iceland the eider ducks are kept in large fenced farms (to protect the birds from predators) with a couple of thousand nesting birds. As it is done at Vega, they prepare nests of dried seaweed for easier cleaning of the down. At some farms they also build simple nesting houses. The manually cleaned down which is produced at Vega is, however, acknowledged to provide much better quality than the mechanical cleaning in Iceland.

The revival of the tradition of producing eider down at Vega was started before the area was designated as a potential World Heritage Site in the mid-1990s. It saw a boost caused by the inclusion on the World Heritage List in 2004. At present there are eighteen people who actively uphold the tradition, compared with six in 2000 (Verdensarvvega.no 2014). Three thousand nests are prepared each year, and a growing number are used by eider ducks. For the last couple of years at Lånan, the site with most eiders as well as property owners present, there have been about 800 nesting eider ducks. This number of birds is close to the numbers that there were in the 1900s. Such figures are motivating for both the property owners on the outer islands who are engaged in reviving the tradition on their sites, and for those in organizational and voluntary work who engage themselves in the support of the revival.

Making sheltered nests for the eider ducks was typical for the tradition of producing eider down at Vegas as in the region of Helgeland. This was not usual further north along Norway's north Atlantic coast. As in the past, the eider duck nesting houses are still most often very simple constructions made of stone slabs, bits of driftwood, parts of metal plates, a plastic or wooden box. An upturned old boat that no longer was seaworthy would usually end up as a nesting house. In fact, people used whatever was at hand. For this building practice, what comes drifting ashore on the tide or in stormy weather has been an immensely important resource. As a result the nesting houses appeared in a great variety of forms and materials. Yet, some types were, and still are, characteristic. Often seen is a saddle-roof formed shelter that could take up to a dozen birds or more. Another common type was constructed by using long planks as base and roof put up against a wall of a building with a row of partitions to make separate spaces for each bird. Previously people also prepared a large number of nests in the boathouse and under buildings for crowds of birds. However, one of the interviewees said that the eider ducks preferred to hatch in shelters made for one bird only. In her experience the birds also preferred the old shelters.

The placement of most nesting houses seems without any apparent order, as they are placed everywhere, scattered in the landscape, on flat ground or in a crop in the terrain, beneath or along the walls of the farmhouses, boatsheds and other human-used buildings, under staircases or the barn ramp. To attract as many eider

Figure 6.3 Nesting houses made for one eider duck each in all kinds of materials scattered throughout the landscape at Muddvær

Source: Photograph by K. Fageraas, Vega, 2013.

ducks as possible people built as many sheltered nests as they could in order to suit most eider ducks' differing needs. In any case, a lot of the shelters would not come into use as a great many eider ducks choose to hatch on the ground in the open landscape. At the sites where down production was an important source of income there could be up to a couple of thousand eider ducks nesting (about the same number as kept on the eider duck farms at Iceland nowadays), which implied a lot of building activity. The varied physical infrastructure of nesting houses provides a distinctive landscape feature.

The human-made shelters have the *quality* of providing protection and comfort to the nesting eider ducks. According to the locals with experiences of the eiders' nesting habits, the human-made shelters were something the birds seem to learn to appreciate. It appears that the younger birds followed the example of the older ones. Usually the eider ducks returned to the same nesting house year after year. But, as one of the interviewees experienced in her effort to re-establish the eider colony, the eider ducks were at first unwilling to hatch in the shelters made for them (I 12, 2008). During the first summer season only one eider duck made use of a nesting house by its own impetus. The interviewee had started by, after a while, placing the shelters over some of the nesting birds. The one bird that by herself, hesitantly, waggled into a nesting house, appeared to be rather old. It had probably been used to hatching in human-made shelters in the past. After 16 years the interviewee had attracted only about 50 eider ducks that hatched in the sheltered nests. So, the unique interspecies relationship of humans and eiders with regard to housing is thus not given once and for all.

Semi-Domestic Bird Species

The making of the nesting houses was part of a traditional economic activity aimed at producing as much eider down as possible to the best quality. To achieve the highest down quality, it was necessary to attract the eiders to hatch in the man-made shelters with nests of dry seaweed, as it was almost impossible to completely clean the down from eiders that chose to hatch in the open landscape (locally called "wild eiders"). Otherwise the down would contain a lot of grass, moss, twigs, sand and other rubbish, thus resulting in lower selling price. The manual cleaning of the down was laborious time-consuming work that went on for months during the summer and most of winter time. It went through several stages. First, the down was sun-dried and the largest bits of rubbish released. Then they would use a wooden frame construction with strings of fishing line to bounce out the smaller bits, before thoroughly handpicking the down to remove every fragment of impurity. According to one of the interviewees it takes about three weeks to thoroughly clean one kilogram of down, the amount needed to make one full-size duvet. On average, an eider duck left less than 20 grams of down in the nest, which means that to make a duvet you would need down from 60 nests. At present about twenty kilograms of eider down are being produced a year at Vega. A duvet costs between 30,000 and 50,000 Norwegian kroner (approx. € 3,500–5,800). There is large price difference between the manufacturers.

The productive aspect of the traditions related to the eiders indicates a form of domestication, which usually is defined as a situation of nonhuman animals being integrated in human socioeconomic organization (Ducos 1978, cited in Ingold 2000, 64). However, such a definition generally implies human appropriation of nature, in which nonhuman animals function as properties that can be inherited and interchanged/sold (Ingold ibid.; Wilkie 2010, 20). Alternatively, a biologically based definition entails a control of breeding to qualify as productive activity (ibid.). None of these criteria are part of the tradition related to eiders in the Vega Archipelago, neither is it just a practice of collecting, as it played an essential part in the human socioeconomic organization of the fisher-farmer's livelihood. Subsequently, the housing of eider ducks and production of down have been termed, by some local and cultural historians, "semi-domestication," first and foremost because of its uncontrolled form of animal keeping (e.g. Næss 2009 [1998], 4; Wold 2003, 125). As it does not involve disciplining actions, such as the control of free range, food and breeding, the production of eider down differs from most other forms of animal husbandry.

From a human perspective, when conceived as a semi-domesticated bird species, the eider duck appears to transgress the human territory, being wild but at the same time seeking proximity to humans. This behavior becomes particularly apparent during seasonal shifts, when the eiders are flocking in fjords or open sea as wild animals in the winter time, while appearing more or less tame during summer when they come close to the human dwellings. Some of the most trustful eider ducks allow people to lift them off the nest to take eggs. This changed nature of the eider duck is fascinating to the island dwellers that experience this transcendence. They talked about them as both timid and trustful: in the Uexküllian ontology the eider duck would appear as neither of these, but primarily as subject to its feral behavior. Despite this, the birds are under some form of human control by being part of, to a large degree, a cultural landscape in which they adapt to people's presence and conduct. Nonetheless, compared to most other situations of animals transgressing the human territory or practices of domestication, the eiders are living as a natural Other, to a large extent unaffected by cultural organization. On closer inspection, the co-existence of humans and eiders reveals, however, that it encompasses a lot more intricate human–bird relations than it first appears.

Adapting to the Habits of Birds

The building of shelters and preparing nests are the only concrete measures made for the eiders by humans. No other form of physical infrastructure was organized. Feeding was not necessary as the eider duck does not take food during the 3–4 weeks when she sits on the eggs. The very nature of the eiders' breeding habits also meant that people had no control over the number of eider ducks that came to their particular island, although the birds usually return to the same nest for years. This uncontrolled breeding and free range of the eider ducks' doings put excessive limitations on the behavior of people who took all possible precautions not to scare the birds. Most of the interviewees in general portrayed the eider duck as easily frightened, and particularly timid in April and May when they walked up on land in the daybreaks to find a suitable nesting place. In this crucial phase people mostly stayed indoors. Cats and dogs were kept away and free-range livestock was moved to islands without nesting eiders. All attention was directed to what suited the eider duck.

One of the interviewees who is still engaged in the tradition of housing eiders was however of the opinion that in the past they had acted really "silly" and thought it was all exaggerated:

> The birds took complete control of your whole life … stupidity! … There were many times we crawled across the floor and we could not walk in front of the windows. Also we hung something for the windows on one side of the house so that the eiders would not be able to see our movements inside. The cows were given food so they would not low early in the morning. Mam did not go to the cow house before ten, eleven o'clock. And we did not use the oven so that there would be no smoke coming out of the chimney. Some were fanatic, absolutely fanatic. And some were less fanatical. To wash clothes you could just forget. I think it was not necessary. But down production was so important at that time. They got more for eggs and down than during a season in Lofoten. Tending eiders was a sacred act.

Other interviewees told me that they were acting in similar ways, but few expressed that it was exaggerated. Rather to the contrary, this way of acting was to a large extent continued today by the practitioners of the tradition and reproduced narratively in the heritage discourse. However, the way people acted in relation to the eiders differed from one island settlement to another. This seems to have been the case in the past as in the present. One of the interviewees said that some were exceptional in this regard: "Yes, out there at [nn] it was very quiet in that period. They hardly showed themselves for the birds. If boats came close to land where the birds nested, they flew off, they were easily frightened. They did almost not see humans."

He himself had experienced that the eider ducks accepted that he had his dog with him when walking around on the island, as well as firing his rifle to scare predators away. The eider ducks next to where he stood took to the wing and flew into the harbor basin, but those nesting a bit further away stayed in the nesting houses. He admitted however that in the past also at his island there were strict restrictions on how people could act when the eider ducks were nesting. As was common, they did not use the oven, he explained, due to the birds as well as all other animals' fear of fire and smoke. They also wore the same clothes so the eider ducks would recognize them. Nevertheless, some of the routines of people's way of cohabiting with eiders seem to have been as much of a distinctive and inherited cultural feature as it was of a necessity to attract and care for the birds.

The different traditions and conceptions of the eider's timidity are likely to be due to differences in economic value it had for the various island settlements. It seems that to those who were most dependent on down production, and with the largest eider population, had imposed the most comprehensive restrictions on their behavior. Another explanation would be that the largest eider colonies probably had a high degree of less timid eider ducks not so much used to human presence. In comparison, eider ducks nesting on the islands which functioned as fishing villages were likely to tolerate a lot of human activity. An interviewee told that the eiders actually moved along with seagulls and other birds when a fishing village at one point was moved to an island further out in the archipelago. Another told that there were never as many eiders at one of the islands as in the year when a massive breakwater was constructed by large, noisy cranes and other machinery. In Iceland, one of the big farms with eider ducks was actually located inside a noisy airport area, where the terrestrial carnivores such as the mink and fox are unlikely to appear (Carlsen 2013, 28). There they also pursued predator control by shooting from hunting towers. To the eider ducks, through the activity, humans were of protection-quality to them. In the opposite case, humans would not be of any functional quality to the birds if they showed no presence or effort to protect them.

A Relationship about More Than Animal Housing

A wide spectrum of relations between humans and eiders was revealed through interviewing property owners from the former inhabited outer islands. Housing the eiders was just one part of this human–bird relationship. Some of the interviewees who were most experienced with the traditions related to the eiders explained that the birds lived off fish waste that people threw into the sea near the shore (eiders are known primarily to eat crustaceans and mollusks, with mussels as their favorite food). One of them said:

> You see, when people lived here, the fish waste was thrown into the sea. It was food for the eider ducks. They were stationary in the harbor area … stayed there all year round, 365 days. But when people moved … the eiders had a different relationship to the site and came back only for breeding. They still do that. But now they don't realize that it is food … when you throw fish waste to them they fly away.

Other interviewees told that the fishermen actually fed the eiders with fish waste. It was not unusual to feed them by hand to prevent the seagulls having it all. One of those who has spent more time than most of the property owners in caring for the nesting eider ducks explained that it was because people gave the eiders fish waste that they became as tame as they were (cf. Suul 2012, 31). According to him the eiders shrieked when they were not given any food by the humans.

This way for the eiders to get food was no longer an option due to the depopulation of the outer islands. In fact, almost all of the fishing activity, which previously was omnipresent in the archipelago, disappeared, due to a collapse in fish stocks and structural changes from coastal fishery by small boats towards larger vessels for deep sea fishing. Toppled by the prohibition by the authorities twenty years ago of throwing fish waste in the harbor basin the eiders had far less favorable living conditions than before in terms of less access to food. Previously, locals had observed that the eiders flocked by the fish landing sites and ate the discards. In 2009 the last fish landing site of the several at Vega was closed down. The bird scientist and ecologist Arne Follestad, who for decades has been observing the birdlife in the Vega Archipelago, supports the locals' experience of the importance of the fishing activity for the eiders presence and numbers in the archipelago (Follestad, cited in Borgan and Alvestad 2012). Since the early 1980s the eider population in the area has dropped by 90 percent; from 30,000 individuals to approximately 3,000 at present (ibid.).

Other causes for the dramatic decrease in the eider population include less predator control. In particular, the human-introduced terrestrial carnivore – the mink – has, according to some of the interviewees, radically decimated the number of eiders. It was introduced in the 1930s, when mink farms were established, a couple of which were located on the islands where the practice of housing eiders was still performed. One interviewee spoke of the mink in an anthropomorphic manner when explaining that it killed with pure delight and could ruin a colony of eiders in one night. Likewise, other interviewees talked about periods with mink present on their islands that killed dozens of the nesting eider ducks. This was the reason that some of them stopped preparing nesting houses to attract the eider ducks. The one-entrance shelters, built for the

Figure 6.4 Nesting houses made of stones and "scrap" is popular amongst the eider ducks. They stand for years, as the eider ducks usually return to the same nesting house year after year. People recognize some of the eider ducks, which are given human names

Source: Photograph by K. Fageraas, Vega, 2009.

protection and comfort for the eider duck, became a death trap if the mink was hunting them. All the dangers to the eiders, as well as humans' capability to work against them, were of major concern to the interviewees who had eider ducks nesting on their islands. Most of them were well aware of the many-sided relations between themselves and the birds, including the feeding of fish waste and predator control.

Heritage Making and the Human–Bird Relationship

In comparison to the knowledge that some local people have about the many-sided relationship between humans and the eiders, a narrow view of this relationship, highlighting, in particular, the practice of animal housing, is dominant within the heritage discourse. There is limited knowledge about the significance of traditional fishing activity, as well as little focus on the execution of predator control, as key factors for the eiders' sustenance and chance of survival. This could be due to the fact that the fishing tradition, in the context of the area's World Heritage significance, was not seen to demonstrate *outstanding universal value*. Neither is the practice of killing carnivores valued much as heritage. The UNESCO advisory expert who evaluated the Vega Archipelago's candidature for inscription on the World Heritage List argued in disfavor of the fishing tradition because it was primarily based on memories (Ogden 2002, 13). It did not represent a living culture, which is a prerequisite for a tradition to be assessed in the context of the 1972 World Heritage Convention (ibid.; see Fageraas 2013, 297). Subsequently, the World Heritage Committee's main argument for awarding World Heritage status to the area was the "practice of eider down harvesting" with its "unique structures" of eider duck nesting houses (WHC 2004). This has imposed a narrow scope both on the fisher-farmer's way of life as well as on the human–eider relationship.

Local enthusiasts, who have been working for the preservation of the traditions related to the eiders, have contributed in bringing the practice of housing eiders to the foreground of the heritage discourse. They have, in particular, been fascinated by the mutual dependence between humans and the eiders that the housing of eiders appears to imply. This focus is also to be found in the nomination file, which they have had the opportunity to contribute to by informing the heritage experts. The concept of "interplay" is much in use to describe how the humans and the eiders relate to each other. However, the nomination file goes further by forwarding the relationship as a conscious reciprocity. It says: "As a form of gratitude for their effort, the people receive eggs and down, and at the same time developed a personal relationship with the wild 'tame' ducks" (MD 2003, 12). In another paragraph the relationship is explained as a "symbiotic contact between man and bird," stating that "both derived value and pleasure from the contact" (ibid., 6). A similar interpretation is equally present in one of few social science papers about the eider-related traditions at Vega, referring to the sociologist Marc Mauss' theory of gift exchange (Sundsvold 2010, 107). Here the breeding time is reflected upon as the key point of exchange, a "reciprocal exchange" based on the "necessary trust" between humans and the eiders (ibid., 97). In such terms a particular anthropomorphic understanding of the human–eider relationship is revealed, and which can be found in all local-based, expert-driven, and scientific discursive fields alike. Based on my reading, this is, thus, far from Uexküllian ontology.

It is likely that those who live close to the eiders on the outer islands have been the primary source for this understanding of the human–eider relationship in the first place. But, if so, it is reproduced and modified in a certain way in the heritage discourse by local enthusiasts and heritage experts. Supposedly, the understanding produced within the heritage discourse resonates back and affects the perception of the island dwellers today. This seems obvious, inasmuch as they talk about the tradition related to the eiders as heritage – but it is not at all as common as I thought it would be, prior to the interviewing. However, they talked a lot about the importance of housing as regards the eider ducks' choosing to hatch in their presence, despite experiencing that the birds would not necessarily use the sheltered nests. Most of the island dwellers had experienced attracting the eiders only by their human presence without making the effort of preparing nests and building nesting houses. In the heritage discourse, the fact that the majority of eider ducks that have been hatching in the open landscape is overshadowed by the focus on housing the birds. The author Inga Næss, who was the first to systematically interview those with experience of the eider-related traditions at Vega, consequently refers to the eider ducks as "farm animals" (see Næss 2009 [1998]; 2013, 148). In this respect, she regards the tradition as a form of domestication. Næss uses, however, the concept of "semi-domestication" (2009 [1998]). This concept is also used (once) in the World Heritage List nomination file, where "domestication" appears most often without a prefix (MD 2003). However, the concept "non-domesticated" is also used in the nomination file. To what degree those who in the past had experience with the eiders' nesting habits thought of their relations with the birds in such terms is not clear. Similarly, my impression is that it also varies among those with eiders nesting on their islands today.

Concluding Remarks

The island realm of the Vega Archipelago is one of the few areas left where the tradition of building nesting houses for the eider ducks continues, although on a scale minimal compared with earlier. This tradition has become subject to heritage production, in particular, through the nomination, and eventually designation, of the area to the World Heritage List. The eider duck nesting houses have become a symbol of the characteristic World Heritage cultural landscape. The main reason for this is that this physical infrastructure is linked to the productive aspect of the traditions related to the eiders, namely the production of eider down. To the World Heritage Committee the so-called "down process" was the essence of the fisher-farmer's sustainable way of life, of which the designation of the area as World Heritage was a celebration (WHC 2004). The sheltered nests were a prerequisite for the production of high-quality down. Such a focus is not at all new. Most records of the eider-related traditions, from the past centuries and way back to the Viking Age, are about the economic aspect of the down product, and the protection by law of those who had the legal right to it. Also, the first ethnological study, by Knut Kolsrud (1975) based on his extensive ethnographic fieldwork in the 1950s, deals mostly with the productive aspect. He terms the human–eider relationship as "eider-breeding," and aims at explaining how different techniques for cleaning the down have been spread and commonly used in various regions. Hence, characterizing the eider ducks as "farm animals" and conceiving their relation to humans as forms of domestication, as in the recent heritage discourse rhetoric, follows a pattern with long historical roots.

In fact, the traditions related to the eiders have much in common with the concept of domestication when it comes to its productive aspect. First, the protective measures taken by humans helped considerably in the successful reproduction of the eiders. In that manner, to some degree, the humans controlled and secured the eiders' breeding. Second, people have provided an excessive supply of food by their fishing activity, which according to interviewees, in turn, has had controlling effects on both the eiders' whereabouts and their tameness. Third, the housing of the eider ducks was part of the socioeconomic organization of the fisher-farmers. It differs, however, from most features characteristic of domestication in that it comprises a much simpler physical infrastructure with relatively few disciplining functions. For example, compared with keeping livestock confined indoors or in fenced areas – as a great deal of modern animal husbandry does – it allows for free range. Moreover, the down was something the eider ducks left in the nest after hatching, as it was plucked from the breast and placed around the eggs to keep them warm. It is not "belongings" taken from the animals while in captivity. In this sense, however, the tradition of housing eider ducks and down production consists mostly of practices of building and collecting.

The heritage discourse has promoted the human–eider relationship as a social contract, which, as I have referred to above, in line with Rhode Wilkie (2010), was typically characterizing the productive relationship between traditional farmers and their animals. Whatever it is called, as a form of domestication or not, and the eiders as tame or wild, the relationship is perceived to be a form of agreement reached as the eider duck changed its behavior from appearing fearful to showing trust in the humans when they demonstrate the ability to provide a protected nesting place. This supposes a kind of shared consciousness, which intersects people and the eiders. Believing in the reality of this interspecies agreement is foregrounding a conception that the eider ducks deliberately choose to enter into animal husbandry. To some island dwellers a number of the eider ducks could be quite tame and tolerant of noisy activity, but they were still, in general, fearful in their behavior. That the eiders would feel pleasure from the contact with humans, which the authors of the nomination file suggest, is uncertain. All such statements, as well as the use of concepts such as "interplay," "reciprocity," "exchange" etc., are interpretations based on anthropomorphism, which goes far beyond the Uexküllian terminology for explaining how animals encounter objects in their surrounding world. They neither represent properly the island dwellers' way of talking about their experiences with, and caring for, the eiders, which was much more plain and simple about both their own human and the eiders' behavior.

The fascination with the eiders' disposition to come close to people and use the human-made sheltered nests has created a dominant narrative about the Vega Archipelago's significance as heritage. It represents, however, a narrow understanding of the human–eider relationship, excluding not least the traditional fishing activity's contribution to the eiders' access to food. Moreover, it overshadows the fishing activity as a central part of local history. This fact was addressed by some of the interviewees who were indignant about the lack of focus on the essential role that fisheries have also had for human settlement and livelihood. When it comes to the number of eiders, the interviewees said it has increased considerably in the archipelago, based on their experience that a lot more eider ducks are coming to nest on their island properties. This development is probably due to far better protection-quality for the eiders as people have shown up again and started to care for them by building more sheltered nests. Hence, the common eider, privileged as it was in the past, experiences the benefits of yet again being privileged through the heritage production that has been going on for the last two decades. The eiders have, for sure, in the encounter with their surroundings, experienced what is a changed context of humans from a marginalized place to an attractive heritage site, resulting in the presence of many more people. To these birds, as for most duck species, what would be an unnatural setting for most wild nonhuman animals, seems to be rather natural to them. That is basically how the human–eider relationship functions at Vega, proving favorable to the eiders as well as fascinating to the humans. With regard to the Uexküllian ontology, however, the human promotion and experience of this relationship – marked as it ought to be by anthropomorphism – seems to emphasize too much a social aspect between the species to explain the eiders' behavior. It might be that domestication is such a dominant way of thinking about other animals in human environments that it has had a strong imprint on the conception of the eiders' animality.

Acknowledgments

This chapter is mainly the result of the interdisciplinary research project *Vega2045* and *Coast-Inland*, funded by the Norwegian Research Council, along with the Strategic Institute Programs at NIKU (Norwegian Institute for Cultural Heritage Research) concerning heritage policy and local effects. I would like to thank the colleagues I've worked with who have contributed substantially to my insight into the nature and culture of Vega, and particularly the project leaders of *Vega2045* and *Coast-Inland* who took charge and led the interviewing during the two main fieldwork periods. Thanks also to my PhD supervisors, and for this chapter in particular Liv Emma Thorsen, who has inspired me as many others approaching the field of animal studies. Also great thanks to the interviewees at Vega who welcomed us to their island properties and revealed to us some of their profound and sophisticated relations to wild birds.

References

Borgan, E. and Alvestad. P.O. (2012) "Ærfugl i fare," *NRK 1-TV* [Norwegian Broadcasting]. Interview with bird scientist Arne Follestad. http://tv.nrk.no/serie/schrodingers-katt/DMPV73002312/13-09-2012#t=19m29s [Accessed 20 May, 2014].

Carlsen, T.H. (2013) *Nordisk ærfugldun. Kunnskapsutveksling mellom Norge, Island, Grønland og Færøyene.* Report. Tjøtta: Bioforsk Nord.

Cassidy, R. and Mullin, M. (eds) (2007) *Where the Wild Things Are Now: Domestication Reconsidered.* Oxford: Berg.

Daston, L. (2005) "Intelligences: Angelic, Animal, Human," in L. Daston and G. Mitman (eds), *Thinking with Animals: New Perspectives on Anthropomorphism.* New York: Columbia University Press, pp. 37–58.

Daston, L. and Mitman, G. (eds) (2005) *Thinking with Animals: New Perspectives on Anthropomorphism.* New York: Columbia University Press.

Fageraas, K. (2013) "Verdensarvens forandringer på Vega: Kulturell verdsetting langs nye og gamle skillelinjer," in G. Swensen and T. Guttormsen (eds), *Å lage kulturminner: Hvordan kulturarv forstås, formes og forvaltes.* Oslo: Novus forlag, pp. 287–334.

Ingold, T. (ed.) (1988) *What Is an Animal?* London: Routledge.

Ingold, T. (2000) *The Perception of the Environment: Essays in Livelihood, Dwelling and Skill.* London: Routledge.

Jørgensen, M.W. and Philips, L. (2002) *Discourse Analysis as Theory and Method.* London: Sage.

Kolsrud, K. (1975) "Dunvær," *Norveg*, 18.

Laclau, E. and Mouffe, C. (2001 [1985]) *Hegemony and Socialist Strategy: Towards a Radical Democratic Politics* (2nd edn). London: Verso.

MD (2003) *Vegaøyan. The Vega Archipelago. Norwegian Nomination 2003 – UNESCO World Heritage List.* Oslo: Ministry of Environment.

Næss, I.E. (2009 [1998]) *Tending the Eider Ducks. Down Islets in Vega.* Gladstad: Vega kommune.

Næss, I.E. (2013) *Øylandet: Dagligliv på Helgelandskysten.* Stamsund: Orkana forlag.

Ogden, P. (2002) *World Heritage Site Nomination. The Vega Archipelago, Helgeland, Norway. Report of Advisory Visit: Peter Ogden. June 2002.* [Unpublished].

Sebeok, T.A. (1988) "Animal in Biological and Semiotic Perspective." in T. Ingold (ed.), *What Is an Animal?* London: Routledge, pp. 63–76.

Sundsvold, B. (2010) "Amenity of Place: Eider Down Harvesting through Changing Time," *Acta Borealis – A Nordic Journal of Circumpolar Societies*, 27(1): 91–115.

Suul, J. (2012) *Edderdun fra nord: Det biologiske, kulturhistoriske og juridiske grunnlaget for fredning av egg- og dunvær samt anvendelsen av andre sjøfugler.* Trondheim: Norsk ornitologisk forening.

Thorsen, L.E. (2006) "Kampen om territoriet: Å skrive dyr inn i kulturhistorien," *Tidsskrift for kulturforskning,* 5(1): 23–38.

Troels-Lund, T. (1968 [1908–1910]) *Dagligt liv i Norden i det sekstende Aarhundrede* (3rd edn). København: Gyldendal.

von Uexküll, J. (2001 [1936]) "An Introduction to Umwelt," *Semiotica*, 134(1–4): 107–110.

von Uexküll, J. (2010 [1940]) *A Foray into the Worlds of Animals and Humans/The Theory of Meaning.* Minneapolis: University of Minnesota Press.

Verdensarvvega.no (2014) "The Egg- and Down-Collecting Tradition." www.verdensarvvega.no/index.php/en/egg-og-duntradisjonen [Accessed 23 October 2014].

WHC (UNESCO/The World Heritage Centre) (2004) *Decisions Adopted at the 28th Session of the World Heritage Committee (Suzhou, 2004).* World Heritage Committee, Original Decision Document. Document: WHC-04/28.COM/26. http://whc.unesco.org/en/decisions/128 [Accessed 4 October 2012].

Wilkie, R.M. (2010) *Livestock/Deadstock. Working with Farm Animals from Birth to Slaughter.* Philadelphia: Temple University Press.

Wold, H.A. (2003). "Det 'resirkulerte' stedet: Vegaøyan – fra utvær til verdensarvsted," in I.M.A. Hauan et al. (ed.), *Karsløy og verden utenfor. Kulturhistoriske perspektiver på nordnorske steder.* Tromsø Museums skrifter XXX. Tromsø: Tromsø Museum, pp. 113–131.

7 Muskox in a Box and Other Tales of Containers as Domesticating Mediators in Animal Relocation

Dolly Jørgensen

In 1929, the ship *Kap Flora I* and her captain Peder Andresen carried a precious cargo to Ålesund, Norway: 26 live muskox calves captured in East Greenland. Captain Andresen had a history of catching muskoxen. Back in 1922, he had successfully brought 11 calves to Norway, which were subsequently sold to American zoos, and in 1925, his crew brought home another eight. His 1929 cargo was the most calves ever captured by one expedition (*Aftenposten* 1929a). As they journeyed over the sea, the homes of the young captives were boxes. It was those boxes that made the relocation of the muskoxen from East Greenland to Norway possible, and in the process, would forever change the animals.

Animal relocation, which is the practice of bringing individuals of a species from their current location to populate another area, depends upon the ability to transport the individuals from point A to point B. Animal relocations may be organized to aid in species conservation (either the animal's original habitat is no longer suitable or the species is being returned to an area in which it used to live) or to benefit humans, which is the case with introductions of game animals and birds as well as fur-bearing animals for harvesting. The logistical and technological challenges of relocating a 200kg muskox across the ocean, or transporting a beaver that chews through wood in a wooden crate, are anything but simple. When scientists discuss animal relocation, they often focus on the ends – the animal released into a new habitat – yet acknowledge that transport stresses can increase mortality (IUCN/SSC 2013). In this chapter, I focus on the means – the boxes and other containers that make that eventual release possible, as well as their ascribed meanings. Unlike many other forms of animal housing, relocation containers are temporary – animals will be in them for hours or days or weeks, but not much longer. In spite of the limited time of containment, these temporary houses are crucial points of transformation in the relocation process.

More than just tools for moving the animals safely, containers change their contents. Rather than serving as an intermediary – a black box that transports meaning without transforming its contents, as defined by Latour (2005) – I will argue that the relocation box creates and shapes the interactions between the animal, humans, and the surroundings, putting the container into the role of mediator. Mediators "transform, translate, distort, and modify the meaning or the elements they are supposed to carry" (Latour 2005, 39). Crucially, the transformation, distortion or modification caused by a mediator can affect both the content and the perception of the content in this definition.

I argue that the boxes used to transport the animals transform their contents on two levels. On the first level, the box makes the animal into a transportable object by *domesticating* it. This objectification process changes the perception of the animal from wild to domestic. During the period of relocation, human caretakers see the animal as a controllable object. From the animal's perspective, the container modifies their relationship to other animals, humans, and their environment. The animals themselves must modify their behavior in order to survive, becoming more controllable, which reinforces the human perception of their domesticity while containerized. On the second level, when the animal emerges from the crate, the human witnesses re-ascribe wildness to it and ascribe belongingness to the wild animal in its new surroundings. When the animal is released from the container, it must reassert its wild nature and fit in with its new surroundings to survive without human intervention. The container transforms the contents from an animal *from somewhere else* to an animal that *belongs here*.

Before exploring these transitions further, I want to acknowledge that the definitions of, and boundaries between, wild and domestic have long been debated (Russell 2002). In general, animal domestication is understood to involve both social and biological components (Clutton-Brock 2012, 3), although scholars tend to emphasize one or the other depending on their perspective. Domestication has also typically been defined as a relationship with a population of animals rather than the relationship between a particular person and a particular animal, which has often been labeled "taming" (Russell 2002). I want to break with that scholarly position and return to the idea of domestication based on its Latin root *domus*. The *domus* is the house, so the Latin verb form *domesticāre* means to dwell in a house or become accustomed to it. The verb *domesticate* at its most simple means "to make, or settle as, a member of a household; to cause to be at home," and more specifically for animals, it is "to accustom to live under the care and near the habitations of man" (OED 2014). This definition is apt in this context, because it means that when an animal is put into a house of whatever kind, it has been domesticated. Within the housing, the animal is living under the care of humans and is necessarily therefore near them. The housing itself as object is integral in animal domestication, even if that domestication is temporary.

How might this type of domestication apply in animal relocation cases? The point of the relocation is to move an animal from one place to another with the intent of releasing it into the wild. When this involves the capture of a wild animal, we might assume that the animal is wild throughout the process, that it began as wild and ends as wild. But I want to show how the box brings about a middle domesticated state, which must be forced upon the animal and then is removed from it.

To examine the box as mediator, I am going to tell three short tales of animal relocation in Norway and Sweden from 1900 to 2013. I have labeled these as "tales" for a very specific reason. Environmental historians are committed to narrative as a form of analysis, as William Cronon so astutely observed over 20 years ago. His observations are worth quoting here: “We configure the events of the past into causal sequences – stories – that order and simplify those events to give them new meanings. We do so because narrative is the chief literary form that tries to find meaning in an overwhelmingly crowded and disordered chronological reality” (Cronon 1992, 1349) In the process of ordering the past into a plot, historians cannot help but judge the consequences of human actions. This has meaning for contemporary readers, not as recovery of “facts” from the past, but rather because in narrating “the dilemmas they faced we discover our own, and at the intersection of the two we locate the moral of the story” (Cronon 1992, 1370). This is what I am trying to do with these tales. It is a very specific form of analysis where I want the reader to open the lid of the box with me to observe its contents, and perhaps even imagine herself inside the box during the relocation process. In telling the tales, we can experience them.

Using examples varied in time and space, I hope to show that container mediation is an inherent part of the relocation process rather than one unique to any given case study. Unlike the other chapters in this volume, the time in this house is limited – it is never intended as a permanent home – yet the container still modifies the relationship between animal and human. In focusing on the container as mediator, I do not want to suggest that other aspects of human–animal relationships, including power relations, naming strategies, concepts of nativeness, and the like are unimportant in relocation. But I believe that by zooming in on the container and the processes immediately surrounding it, we can begin to understand the ways that even temporary housing modifies both human and animal behaviors and thoughts. In my first two tales, human caretakers treat the animals as if they occupy roles in the household: in the first one as a pet, in the second as a child. In the third tale, I explore how an animal which has been born and raised in captivity can be transformed into a wild inhabitant through the relocation crate.

Muskox in a Box

Andresen’s cargo of muskox calves in 1929 had all been caught in the Norwegian territory of East Greenland, where Norwegian seal trappers had been active since the late 1800s. The muskox (*Ovibos moschatus*) is a strange looking creature, something like a cross between a mountain goat and an ox. It is indeed related to sheep and goats, but its specialized horns and propensity to charge when threatened are reminiscent of oxen.

The first muskox calves taken from Greenland had been captured in 1900 under the leadership of zoologist Gustaf Kolthoff. The previous year, the Swedish geologist Alfred Gabriel Nathorst had observed wild muskox herds in Greenland during an expedition and came to the conclusion that the Swedes should try to import and domesticate the animals (Nathorst 1900). The main purpose of such an endeavor would be to corner the market in muskox wool, considered one of the highest quality wools available. Nathorst believed that the muskox would acclimatize perfectly to Sweden and be even more productive than reindeer, since the animal was more tolerant of mosquitos, less prone to wolf predation, and did not require long-distance migrations over the year to new feeding grounds. On the recommendation of Nathorst, Kolthoff's expedition had first tried capturing adults, but after little success, decided to capture calves. Since the calves were well-protected by the herd adults, the only way to capture the calves was to kill the adults in the herd first (Kolthoff 1901).

Kolthoff offered a justification for the decision: "Surely two muskox calves in Sweden are much more valuable than six in Greenland" (Kolthoff 1901, 178). The value of the animal according to Kolthoff depended on its geographical location. Moving a muskox from one place to another was understood as changing its value – it would be worth more in Sweden than it was in Greenland because it could be directly used by humans. The box that transferred the calf from one locale to another thus modified its value.

Eventually the expedition successfully captured a male calf, which was given the name Hjalmar. Kolthoff described the scene when the calf was brought onboard:

> He was put into the crate built for a muskox cow and his bands [around his feet] were loosened. At the beginning he was very angry, running toward whoever came near to him and took out his wrath on a hay sack, which he like a ram ran incessantly against so that the sack took an enormous shock from his forehead. Very soon, however, he became calmer, and after a few hours he started to eat.
>
> (Kolthoff 1901, 193)

After the caging and taming process, Kolthoff claimed that Hjalmar was "as a tame as a dog. He followed after us over the whole deck and liked it best in the kitchen" (1901, 203). This comparison of muskox and dog is key. Hjalmar had been brought onto the ship and put into the crate. When that happened, he was moved into the domestic sphere – he was completely dependent upon his human caretakers. In spite of Hjalmar's initial objection, he learned to fit into the household and took the role of pet, "as a dog."

Andresen's later experience with capturing and crating muskoxen over the sea was similar to Kolthoff's. When he was asked if the muskox calves were difficult cargo, he replied:

> Oh no, they are very decent, muskoxen, – you only have to become familiar with them. We have them in spacious, sturdy crates made of wooden planks. To start with the animals try their strength on the containers, but when after a while they realize that the crate is stronger than themselves, then they take it easy. After one or two days of course they begin to take food. … Muskoxen soon become friends with humans. After they have been in their crates on deck for 8 days, they recognize the goodies that they get when the boys approach them with grass or moss. The youngest among the animals are easiest to acclimate – we therefore capture the youngest animals possible.
>
> (*Aftenposten* 1929a, 2)

The calves in Andresen's description are similar to Hjalmar. They become "friends with humans" and recognize treats. These are characteristics that would generally be ascribed to pets.

Photographs from another expedition in 1929 show the crating process in action (Figure 7.1). The *Veslekari* was commissioned for scientific expedition to East Greenland by Docent Adolf Hoel of the *Norges Svalbard- og Ishavsundersøkelser*, which later became the Norwegian Polar Institute. Hoel combined the eight calves captured by the *Veslekari* crew in 1929 with 10 bought from Andresen's 1929 expedition and then released them together as a herd on Svalbard, an Arctic archipelago under Norwegian sovereignty. Hoel intended the muskox herd to grow and become an economic force on Svalbard as a source of meat and wool (*Aftenposten* 1929b). Although the muskoxen would be free-roaming wild animals, the vision was that they could be hunted for meat and their wool could still be collected. When the calves were released on Svalbard, they emerged from their wooden crates but instead of the calves running away into the wild, they approached the man standing in front of the crates, probably looking for a treat to eat like the ones they had received while inside the container.

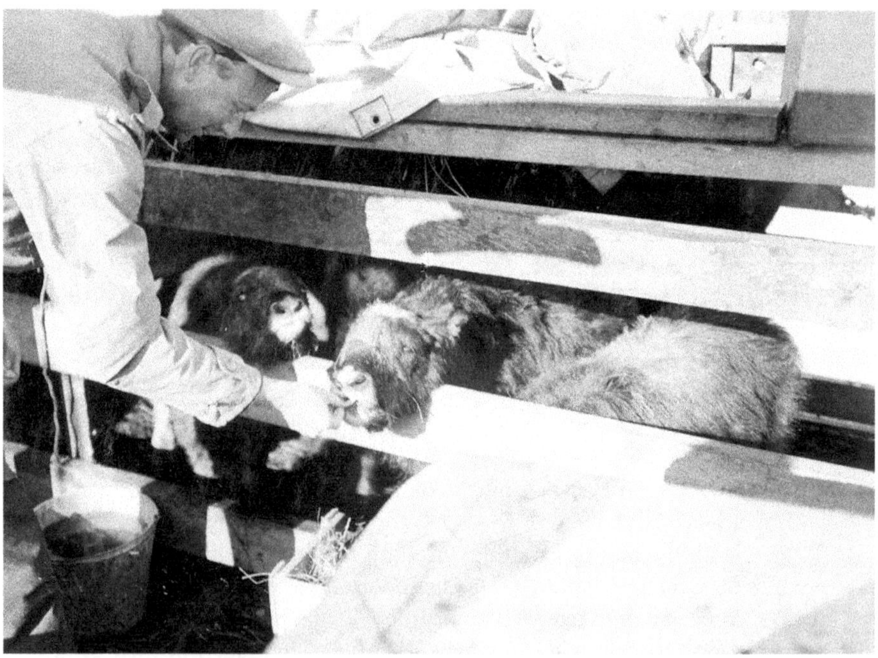

Figure 7.1 Crew member Luncke feeds muskox calves onboard the *Veslekari* on 26 August 1929

Source: Photograph by Andres Kristian Orvin, Norsk Polarinstitutt archive, NP041421. Reproduced with permission from Norsk Polarinstitutt.

The narratives of the captains and the photographs stress the placement of the muskoxen into the crates as the first step in their transformation. The container mediated the relationship between human and animal on the voyage. The calves adjusted their behavior after entering the crate, eventually calming as they became resigned to their fate in captivity. After some time crated, the calves would allow more and more human interaction, eventually even desiring it. The humans understood the crate as the instrument of domestication during the relocation process. The crate turned the muskox from wild creature into domesticated pet.

In the narrative of relocating muskoxen from Greenland to Scandinavia, the crating process is a domestication process. The calves abused the sturdy wooden crates as they tried to escape back to the wild, but upon realizing the futility of their wild behavior, they became tame. As a strategy for relocating animals that will eventually be released back into the wild, the domestication process could have proved counterproductive. This possibility was even recognized at the time by Hoel, who was worried that the animals might make their way to the nearest village for food in the winter since "they have become so used to people now" (*Aftenposten* 1929c). Yet it does not appear that Hoel's warnings came true, as wandering muskoxen on Svalbard did not become an urban nuisance until the 1950s. This means that after the animals were released from their crates and the men left the scene, the muskoxen reverted to their wild behaviors. From the point of view of the animal, the transformation of the containers' contents had only been temporary.

From the point of view of the humans, however, the container transformed a wild animal in Greenland with little value into a potential resource on Svalbard. It turns out that the resource was never actually harvested – muskoxen were always a protected species on Svalbard, so it was illegal to hunt them and wool was never collected in any great amount given the terrain and the muskox's propensity to range far and wide. Those later developments were not, however, known to those involved in the transplantation process in the years around 1929. To them, moving a muskox in a box changed its meaning and value.

Boxed-in Beavers

When the first reintroduced European beavers (*Castor fiber*) came to Sweden, they had to take a long and arduous journey to their new home. First, they had to be moved from their capture site near Åmli, Norway, to the train station in Oslo, then onward via train to Stockholm in the autumn of 1921. After wintering in Skansen zoological garden in Stockholm, they travelled north on the overnight train on 2 July 1922. In Östersund, they had a break and got on another train to Strömsund. After a steam boat, a horse carriage, car trip, motor boat, horse sled, and boat again, the little beavers were finally at the release site – a whopping four days after leaving Skansen in Stockholm (Arbman 1922; Festin 1922). The difficulties of getting the beavers from their original home in southern Norway to a release site in mid-Sweden made transportation technologies absolutely critical to the success of the project. On 6 July 1922, they reached their destination of Bjurälvdalen in Jämtland and were released into the wild. They were the first beavers in Sweden since 1871, when the last Swedish beaver had been seen.

During the relocation, the beavers were crated in a wooden box. The same box was used for the entire journey, from the time they left their original home in southern Norway to their new home in mid-Sweden (Arbman 1922). Nils Thomasson, a professional photographer, documented the trip from Strömsund to the release point, a mode of recording the reintroduction history for the future. In the process of documenting the beavers' journey, Thomasson's photos, which are kept at the Jamtli archive in Östersund, almost exclusively show the beaver box rather than the beavers themselves: the box is shown on a horse cart (Figure 7.2), being carried on poles, in a boat, and being pulled on a horse sled.

There is only one extant picture of the beavers themselves. In that photograph, the beavers are being examined by Sven Arbman, a zoologist who was responsible for the beavers' health during the journey (Figure 7.3). Even in this case, the beavers are contained within the box, which takes up much of the image. As Eric Festin (1922, 90), who had organized the relocation project noted: "For 3 ½ days,

Figure 7.2 During the journey of the first beavers bound for reintroduction in Sweden, the box containing the beavers was put on top of hay bales to try to cushion the ride

Source: Photograph by Nils Thomasson, 1922, Jamtli archive NTh621. Reproduced with permission from the Jamtli archive.

Figure 7.3 Sven Arbman checking the condition of the beavers on the last day of their journey

Source: Photograph by Nils Thomasson, 1922, Jamtli archive, NTh626. Reproduced with permission from the Jamtli archive.

the beavers had stayed in their crate." The box in the photos essentially takes the place of the beavers. Viewers see the box, but they are intended to think about its contents.

The box, however, was more than transportation – it was transformation. The relationship between human and beaver in this episode is founded on the *domus*, the house/household. The box transformed the beavers temporarily into domesticated wards, just as it had with the muskoxen. Arbman recorded how the box had to be moved from one train to another by strong railroad workers. The beaver pair had to be temporarily removed from the box so it could be cleaned after the first overnight train.

> When the lid was lifted, they rolled into balls and buried their heads together in a corner and let out a faint whimpering like little children, the only sounds their vocal organs could produce. They seem to me good-natured, phlegmatic without strong emotions. The Norwegian trapper's depiction of a wild animal

> that sits on its hind legs and gnashes his teeth so that one "cannot come near" does not fit with the animals he delivered us.
>
> (Arbman 1922, 274)

In Arbman's description, the beaver are portrayed as children. Arbman interpreted the beaver behavior while in the box as docile and childlike. They did not behave like wild animals that would bite people who came too near.

The crate transformed the beaver from wild animal into a domesticated one. This animal needed care and attention. Fresh water, bark, and twigs were given to the beavers in the box along the way. Arbman (1922) was concerned about the beavers' condition and regularly monitored the beavers' temperature by touching their noses to determine if they were too hot, and checked their general health condition (as in Figure 7.3). When Festin (1922) reported on the reintroduction, he commented on the difficult journey and that some of the choices, like putting the beaver box on top of hay sacks for cushioning, would not be repeated. The bumpy journey by horse had given these first beavers "a fever", luckily cured by the tranquil boat ride that followed. The beavers in the box had to be taken care of like sick children.

When the beaver in the box took on the role of child, humans became like parents. The beavers carried in the box to Sweden had originally been caught by Peder M. Jensen-Tveit of Åmli, Norway, which featured the last holdout population of beavers in the country. According to his obituary many years later, Jensen had familial feelings about the Norwegian beavers he captured and released in far away places:

> Typical for his attitude to animals, to all animals, he followed the beaver transport to the release location. "They must have the last goodbye in Norwegian when they are set free a long way from home," he said and patted the animal like a father on the neck. This laying on of the hands would have resulted in one hand less had it been anyone other than Beaver-Jensen, but he and the freshly caught beavers understood each other.
>
> (Tjomsland 1963, 11)

In this passage, Jensen, who caught the beavers in the wild and then facilitated their relocation and release, is a father, which makes the beavers his children. On some of Jensen's business letterhead, a series of pictures of his operations includes one of him feeding a baby beaver with a bottle. The image places Jensen as parent and beaver as child. Those beavers needed those last words whispered in Norwegian to them to remind them of who they were and were they were from – what family they belonged to – because now they would take on a new identity and leave the family.

The same kind of integration of beaver into the domestic sphere occurred in the second beaver reintroduction in Sweden, which took place in 1924 in the northern country of Västerbotten where a county hunting association took the lead in bringing beavers back. In texts published in the association's yearbook, the relationship between the hunters and beavers is not described in terms of conquest or domination, but rather of co-citizens and family. One photograph shows a member of the reintroduction group with a beaver captioned as "State forester Johansson cheering up one of his new 'countrymen'" (Anderson 1925). A long article on the reintroduction of beavers begins by listing the "baptism officiant" (*dopförrättare*) and "godparents" or "sponsors" (*faddrar*) for the reintroduced beavers (Wahlberg 1925). The use of these titles signals that these were more than people who observed the reintroduction – they are the people who became responsible for the beaver's success. Just as a baptismal sponsor agrees to raise a child to know God and the church, these individuals were agreeing to ensure the beavers' integration into the landscape. Sponsors are not in a position of domination, but rather facilitation and guidance. In fact, an article in 1930 refers to the beavers as the "wards" of Lennart Wahlberg, who was listed as the "officiant" in the release of the four beavers on 20 July 1924 (von Post 1930).

This wardship was, however, temporary. After the beavers were released, there was no invasive management of them. There were regular "beaver reports" on a monthly or quarterly basis to document the beavers' success in their new homes, but there was no direct conservation intervention such as relocation or supplemental feeding. The children grew up quickly.

The box also served as a second type of mediator in the reintroduction process: the beavers, which were captured in the wild *in Norway*, were reconfigured into wild beavers *in Sweden*, with a brief respite as domesticated in between. According to the sentiments of those involved in the relocations, when the beavers entered the box they had been Norwegian, but when they left it, they were Swedish. At that moment in which the beavers were "lifted gently out of the box and set on the shore" they were "reincorporated into the Swedish fauna" (Arbman 1922). The box served to transform the beaver from outsider to insider in the Swedish landscape. To bring beavers back to Sweden was nothing short of a revolution, according to Arbman:

> We celebrate the solemn moment at 3.30 am when the beaver was reincorporated into the Swedish fauna, when we got to feel with our hands the revolution in which we participate, the prank which seems to turn everything upside down, the "order of nature," the "march of progress" and all that it entails, that mankind goes forth and multiplies and takes possession of the earth and self-evidently eradicates all other creatures if they are not good enough as slaves.
> (Arbman 1922, 278)

The box, as a tool in the hand of humans, had re-ordered the relationship between people and nature through its ability to transform the beaver from Norwegian-other to Swedish-native. This was a transformation in the way that humans understood the animal, rather than a modification of the animal itself. The box transformed the relationship between human and animal by making the beaver that exited the box into a Swedish resident at home here rather than a Norwegian resident that belonged there.

The temporary container resulted in only a temporary domestication, but a critical one for making the relationship between human and animal during relocation work. The container mediated the animal housed within it, causing the animal's temporary transformation into a child to be cared for. The beaver within the box existed in a liminal state between being wild and being wild again. In fact, in order to succeed and reproduce, the relocated beavers *needed* to become wild and embrace their new homes. They could not stay in a domestic state. When the animal left the container for good upon its release, it was de-domesticated – it became wild again in its new home.

A Flying Muskox

While the muskox can catch an onlooker off-guard with quick bursts of speed, it cannot fly, except with the help of humans. On 4 April 2013, a two-year-old muskox named Idun, who was born and raised at the Myskoxcentrum breeding center in Tännäs, Sweden, was drugged and then put into a crate only slightly bigger than herself. The crate was subsequently attached to a helicopter and flown up into the mountains near where the small Swedish muskox herd of six individuals had been spotted. The doors on either end were opened up and Idun took her first steps as a free-range muskox in the Swedish mountains.

This chain of events was documented in both video and still photography, made available for public consumption on the internet, including news broadcasts up to one month later (e.g. SVT 2013; NRK 2013). Myskoxcentrum announced Idun's release with a group of photographs posted 5 April 2013 on its Facebook page (https://www.facebook.com/myskocentrum). The set of eight images is striking in its presentation of the transportation box in six images, while only four show the muskox itself. The first three photos show the helicopter with a box hanging beneath it. In these, the muskox is not directly visible, but the muskox is not missing from the images. Instead, the box acts as a replacement for the muskox – when the viewer sees the box, she is supposed to see the muskox. This becomes more readily apparent when the helicopter images are seen in conjunction with the next two images of Idun leaving the box. Here, the muskox emerges from the box physically. After a glimpse of Idun scurrying away, the box is shown to the viewer empty in photo 7, followed by a view of Idun in the distance in photo 8.

In the photographic series, the box not only transports, but also transforms its contents. When the muskox is in the box, it is in a liminal state. Idun entered the box as a domestic muskox who had daily contact with humans who fed her hay and managed her enclosure. She was on display at the Center like a zoo animal that visitors come to photograph. During her transport, she disappears, being replaced by the box. Then finally the box stands empty as the muskox has been transformed from domesticated animal to wild one. Idun had never been wild like muskoxen taken in boxes across the ocean from East Greenland to Svalbard, yet like them she became wild when she was released on the mountain.

Idun's transformation into wild muskox is a small part of a larger story about how muskoxen came to be in Sweden and how people would respond to their presence. Muskoxen had not been in the mountains of Härjedalen, Sweden, for thousands of years before a small flock of six animals moved there in 1971 from the Dovre mountains of Norway, where muskoxen had been reintroduced earlier in the twentieth century. The flock established themselves as border-crossers who migrated between Norway and Sweden through the mountains of Femundsmarka and Rogen. The group reached a high point of 34 animals in the mid-1980s, but the population went sharply down afterward (Lundh 1992).

The population crash prompted some concern. Four representatives of the Green Party (*Miljöpartiet*) in the Swedish parliament proposed a motion in 1991 that the government should work toward saving the muskox herd (Riksdag Motion 1990/91:Jo759). The motion suggested that wild muskoxen might be given supplemental food, more animals to reinforce the population could be purchased and released, and muskox conservation could be started in cooperation with Norway. The motion was subsequently dismissed because the committee handling the environmental plan under which the motion was made felt that muskoxen, while important to the tourism industry, were not natural in Sweden: "Muskoxen have been extinct for such a long time in Scandinavia that they can no longer be considered as part of our natural fauna. … A strengthening of the herd would according to the experts become costly and cumbersome and take a disproportionate share of the available conservation funds" (Riksdag Motion 1990/91:JoU30).

Not to be dissuaded, three Green Party representatives filed a motion again in 1999/2000 requesting an official statement from the government on the protection on the muskox population (Riksdag Motion 1999/2000:MJ763). The answer from the environment minister Kjell Larsson once again reiterated the "newness" of the muskox in Sweden but pointed to current work by the Swedish EPA (*Naturvårdsverket*) on a muskox conservation plan (Riksdag Svar på skriftlig fråga 2000/01:1138). The conservation plan, however, was never approved because the muskox was not listed as a native species on the Swedish Red List.

The herd of muskox in Sweden thus stands at a fascinating junction of belonging/not belonging. The species does not have a conservation plan in force in Sweden and it is not recognized as eligible for listing as an endangered species in Sweden because it is classified as "introduced" (Gärdenfors 2010). In this sense, the official line is that muskox does not belong in Sweden. At the same time, there has been a concerted effort to establish a breeding center in Härjedalen with the express intent of strengthening the wild muskox population within the country. Idun's translocation from the breeding center to the Swedish mountains was the first time a muskox was intentionally released to the wild in Sweden.

Seen in this light, the box that transported Idun to the snow-covered mountain becomes all the more transformative as a Latourian mediator. Before the box was flown by helicopter onto the mountain, the muskox herd's right to be there might have been questionable. The human hand helping the muskoxen was limited to a domestic herd on a few fenced-in acres in Tännäs. Upon leaving the box, Idun became part of a wild herd of muskox inhabiting the Swedish mountains with a presumed right to do so. She embodied the conservationists' hopes and dreams of a healthy Swedish muskox population. As such Idun had to shed her domestic ways and become a part of the wild herd instantaneously upon leaving the box.

Telling Tales of Containerization

In each of these short tales of animal relocation, the container – the box which nominally takes the animal from its prior home to its new one – is more than a physical transport method. I have argued that the animal within the housing is *domesticated* – not in the sense of having been bred by humans for a purpose, but rather in the sense of being within a house under the care of people. It is the container itself that facilitates a change in the animal and its relationship with people. The box makes the animal dependent upon human caretakers. From the human perspective, this can place the containerized animal into the role of family member or pet. The container used in relocation acts as a mediator that transforms the animal into a domesticated member of the *domus*.

The containerization is, however, a temporary state. Because the container is not designed as permanent home, the domestication is also transitory. In each of these tales, the animal ends up in the wild. The route to wildness after relocation is always through domestication, whether the domestication had been existing since birth, as with Idun the muskox, or was a temporary condition like the muskox calves from Greenland and the beavers from Norway. In either case, the animal will eventually leave the box and its domestic state. The humans involved in the reintroduction consider the animal "wild" as soon as it leaves the box and assume that it "belongs" to its new surroundings.

The box is transformative, modifying its contents and the perception of them. On the first level, the box works because the animal inside is domesticated. If the animal continued to believe it was wild while it was in the box, it would kill itself by continuing to pound against the side of the crate. Instead, like the muskox calves transported from Greenland, the animal eventually gives in and resigns itself to its domesticated status. During crating, the humans have the responsibility to provide food, water, and protection for the animal. On the second level, the relocation would not work if the animal was not able to transform itself to be wild upon leaving the box. The people need to believe this too – the final 'wild' status of the animal is why we are doing it – and the humans have to accept that the animal belongs in the new place. Wildness in this sense is a counterpoint to domestication. A wild animal finds its own food and shelter and finds other animals with which to be in a herd or flock or pair. These are the criteria for success in the wild, since the animal will no longer be in direct human care within the *domus*. The muskox in a box must learn to live outside the box.

Peering into the containers in these tales allows us to see that the work performed by the crate as technological artifact in animal relocation reaches far beyond the logistical. The moral of these stories is that even a temporary housing structure modifies the relation between human and animal. The power of technologies of simple construction, like the wooden crates or wire boxes in these examples, to re-order relations should not be overlooked. As I have argued elsewhere, simplicity of design does not equate to lack of social complexity (Jørgensen 2008). These are simple wooden or metal crates that would not attract attention as design objects; yet, as objects that order relationships, they play a central role. More than moving the animal from point A to B, the box serves as the critical site of transformation through its ability to domesticate the animal contained within and its ability to "wild" the animal upon it leaving.

Acknowledgments

This research was funded by the Swedish research council Formas through a Young Researcher Grant to the author for the project "The Return of Native Nordic Fauna." All translations are by the author.

References

Aftenposten (1929a) "Vil moskusoksen trives på Svalbard?," August 27, morning edition, p. 2.

Aftenposten (1929b) "Moskusokser på Svalbard vil bety en verdifull forøkelse av ø-gruppens viltbestand," August 27, evening edition, p. 1.

Aftenposten (1929c) "De første moskusokser til Svalbard," September 11, p. 1.

Anderson, A. (1925) "Då bävern återbördades till Västerbotten," in *Västerbotten: Västerbottens Läns Hembygdsförenings Årsbok 1924–1925*, 282–286. Umeå: Aktiebolaget Nyheternas Tryckeri.

Arbman, S. (1922) "När bäfvern återinfördes i Bjurälfven," *Svenska Jägareförbundets Tidskrift*, 60: 274–280.

Clutton-Brock, J. (2012). *Animals as Domesticates: A World View through History*. East Lansing: Michigan State University Press.

Cronon, W. (1992) "A Place for Stories: Nature, History, and Narrative," *The Journal of American History*, 78: 1347–1376.

Festin, E. (1922) "Bäverns återinplantering," *Jämten*, 16: 84–91.

Gärdenfors, U. (ed.) (2010) *Rödlistade arter i Sverige 2010* [The 2010 Red List of Swedish Species]. Uppsala: ArtDatabanken, SLU.

IUCN/SSC (2013) *Guidelines for Reintroductions and Other Conservation Translocations*. Gland: IUCN Species Survival Commission.

Jørgensen, D. (2008) "Cooperative Sanitation: Managing Streets and Gutters in Late Medieval England and Scandinavia," *Technology and Culture*, 49: 547–567.

Kolthoff, G. (1901) *Till Spetsbergen och Nordöstra Grönland år 1900*. Stockholm: Fr. Skoglunds Förlag.

Latour, B. (2005) *Reassembling the Social: An Introduction to Actor-Network-Theory*. Oxford: Oxford University Press.

Lundh, N.G. (1992) *Muskoxe*. Östersund and Trondheim: Länsstyrelsen i Jämtlands Län / Flykesmannen i Sør-Trøndelag / Sør-Trøndelag Fylkeskommune.

Nathorst, A. (1900) *Två Somrar i Norra Ishafvet, Första delen: Kung Karls land, Spetsbergens kringsegling*. Stockholm: Beijers Bokförlagesaktiebolag.

NRK (2013) "Svensk innsats for å redde moskusstammen," *Nyheter*, Broadcast 29 April. www.nrk.no/video/svensk_innsats_for_a_redde_moskusstammen/68435B071CC30437/.

OED (Oxford English Dictionary Online). 2014 (March). "domesticate, v." www.oed.com/view/Entry/56668.

Russell, N. (2002) "The Wild Side of Animal Domestication," *Society & Animals*, 10(3): 285–302.
SVT (2013) "Myskoxe på grönbete," SVT Regional Västernorrland News, Broadcast 4 April.
Tjomsland, A. (1963) "Dødsfall," *Aftenposten*, 1 October, p. 11.
von Post, G.H. (1930) "På besök hos Västerbottens bävrar," in *Västerbottens Läns Jaktvårdsförenings Årsbok 1930*. Umeå: Västerbottens Läns Jaktvårdsförening, pp. 49–66.
Wahlberg, L. (1925) "Bäverns återbördande till Syd-Lappland," in *Västerbottens Läns Jaktvårdsförenings Årsbok 1925*. Umeå: Västerbottens Läns Jaktvårdsförening, pp. 6–23.

8 How Much Is That Doggy in the Window?

The Aesthetics of Shelter Animal Display

Irene Gustafson

Décor and Its Narratives

The German-born film director Douglas Sirk was frequently referred to as the "director of décor". In his1950s era Hollywood films, the sets, props, lighting, and colors, as much as anything else, narrate the story – the architecture directs the characters through the narrative; the wallpaper holds and reveals, in its texture and shape, their deepest feelings. As Laura Mulvey has written about Sirk's *Written On The Wind* (1956), "For he (Sirk) is both a supreme stylist of the cinema, and one whose mise en scène is always in touch with his characters, carrying in image the emotion they fail to articulate in so many words" (Mulvey 2001). Mulvey's description, particularly her use of the phrase "in touch," gets to the heart of the matter. Sirk's films are object lessons in the power of objects. His films carefully image the intimate and tactile presence of the material world and its sticky relationship to ideas, values, and affects. Not in the way that objects or structures are bearers of singular meanings, but rather in how they are always present in the drama of experience: objects orient us toward the world; their very "thingness" stages an encounter and solicits a response. Sarah Ahmed has described the importance of objects this way:

> If consciousness is about how we perceive the world around us, then consciousness is also embodied, sensitive, and situated. This thesis does not simply function as a general thesis, but can also help show us how bodies are directed in some ways and not others, as a way of inhabiting or dwelling in the world .... The objects that we direct our attention toward reveal the direction we have taken in life. Other objects, and indeed spaces, are relegated to the background; they are only ever co-perceived.
>
> (Ahmed 2006, 27)

For Sirk, objects and architecture are the mercurial figurations of the ideological forces that shape the world of the narrative: class, gender, race, and sexuality. Built environments turn the characters towards and away from each other, objects authorize their impulses towards disavowal, aggression and misrecognition and, on rare occasions, are the surfaces of insight and revelation. The politics of his cinema, it could be argued, was located in the traffic between on- and off-screen space – his cinematically stylized worlds presented to the viewer as an uncanny version of our own, an imitation of the life we are already living and, like his characters, is one that we most often fail to articulate in words. His films remind us to be sensitive to the ways in which our perceptions of the world are always and already situated and embodied. Or, to rephrase this with a slightly different emphasis: our perceptions arise within the unfolding drama of bodies inside situations.

This subject of this writing is décor and its narratives. The object of my study is a discrete set of architectural spaces – animal shelters built in the San Francisco bay area between 1991 and the present.[1] As my introduction suggests, my analysis is concerned with aesthetics and its relationship to experience. Aesthetics here refers not only to a style of architecture, but also to the specific capacities for doing, feeling, and thinking that these spaces accommodate. Indeed, what these spaces even allow us to do – how they direct us in movement and action. As my site of interest is one where encounters between people and animals are engineered to occur, my analysis pairs animal studies with visual culture theory, broadly defined, as each of these approaches allows some facet of my topic to appear. These theoretical tools animate my analysis, but a set of materials constitute its shape – my research comes primarily from archival sources (institutional archives, both paper and audiovisual), from an interview with a shelter architect, George Miers, and from time spent inside these shelters experiencing, observing and photographing them. Throughout this piece of writing I will argue that these sites are culturally and politically significant, not because they are extraordinary, but rather because they reveal something fundamental about our ordinary lives. They are sites of ordinary and everyday processes of what we might call domestication. Domestication is defined here in two distinct but deeply interrelated ways: first, domestication as an articulation of the domestic sphere, the private, the interior, the familial; and second, domestication as a relationship with animals, conceived here not as an a priori condition, but as ongoing and processural. This ongoing relationship may be "exploitative or mutual, intentional or serendipitous," but it is nonetheless a doing, thinking and feeling with animals (Cassidy 2007, 12). Domestication is a kind of world-making with animals or, in this case, world *building*. These shelters uniquely stage, in aesthetic terms, domestication as a style, ideology, and a set of experiences. Cats and dogs are our close companions and are increasingly both the subjects and objects of the cultural, political, and affective processes that constitute the terrain of the private and domestic spheres – home, identity, belonging. What follows is not an exhaustive account of these newly-built shelters but is instead a close reading of several buildings where "homeless companion animals" (as shelter dogs and cats are referred to) are housed.

To enter these buildings is to be enveloped within a pastiche of architectural citation, reminiscent of a place that you've been in before: department store, shopping mall, community center, living room, and zoo. If you spend enough

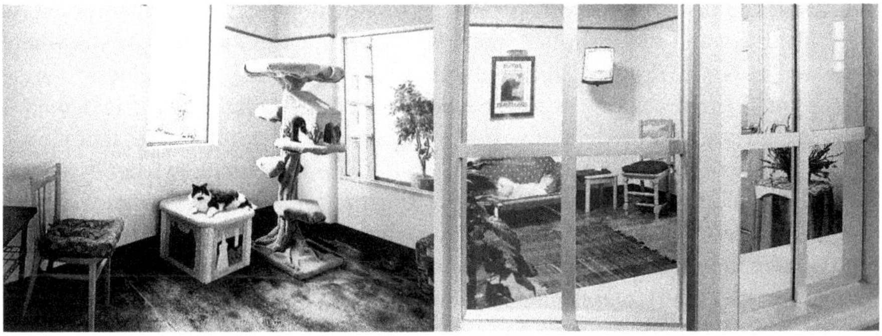

Figure 8.1 The San Francisco SPCA. circa 1998

Source: (left) © ARQ Architects – Architect, © Bob Swanson/Swanson Images – Photographer; (right) © ARQ Architects – Architect, © Jane Lidz, 1998 – Photographer.

time in these spaces, after you've settled in and have processed the signifying systems at work around you, questions will most probably arise: What captures my attention? Where can I walk and what can I see? What does this space feel like? How do I feel here? What types of encounters are encouraged or even possible? If one is willing to discipline these smaller experiential questions into something like a unifying inquiry, one might ask: how do these shelters both produce and reflect the exquisitely complex intricacies of our ongoing relationships with companion animals? "Companion animal" is, of course, a term we use to specify a particular human/nonhuman relationship, a term that refers to the emergence of creatures who are "not meat animals, are not lab animals, are not wilderness animals, are not war dogs, not vermin," which are part of a very particular ongoing intimate historical relationship between people and creatures, genetics and culture, discourses and bodies.[2] Dogs and cats are creatures we've been thinking and feeling with for a long time. They are characters in our narratives of progress but, more than that, they are agents of interaction, actors in entangled networks of nature-culture. And, importantly, they are creatures with which we make a home.

A Brand New Concept in Sheltering

While each shelter, each community, and each individual animal welfare agency is significantly unique, there is a recognizable architectural genre that emerges when these newly-built animal shelters are viewed in aggregate: bright and modern, situated near other public and commercial buildings, they are designed to accommodate the complex bureaucracy of contemporary companion animal welfare – education, outreach and advocacy, fundraising, spay/neuter and veterinary services, rescue, adoption, and euthanasia. The most persistently recognizable features of these spaces are the "habitats" – those areas where the public encounter adoptable animals and where the dogs and cats live. Fronted by large display windows, visitors to the space are invited to gaze into rooms where dogs and cats are featured in home-like settings. Lauded as "cageless," these spaces most often feature exposed wood and tile, furniture and televisions – domestic props designed to invest the scene with a comfortable familiarity. Animal shelters are rhetorically dense sites, designed to articulate what we imagine a companion animal is or should be: a creature of *home*.

Home is by no means an easy or transparent concept. It is one that carries within it complex notions of inside and outside (species, race, class, gender, nation, kin and kind), and is charged with the affective energies of belonging or not-belonging. Home is a verb and noun, and as such the buildings themselves bear the trace – in form and content, structure and administration – of our often messy and always historically inflected conceptions of home. The recuperative work of the shelter is to transition its animal occupants from their status of being ontologically outside – "homeless" – to the status of being firmly situated within the confines of home's literal and discursive boundaries. Architects like to say that buildings are narratives, and certainly, thick narratives of domesticity and domestication, companionship and commodification, abandonment and rescue, set the foundation for these sites and moisten the air that everyone and everything breathes. If a narrative requires a structure, a conflict to be resolved, something that propels it forward towards its denouement, the overarching narrative drive within the animal shelter is the tension between "finding a forever home" (as adoption is referred to) or euthanasia. Home – as a literal architectural space, a concept, a set of institutional practices, and even the redemptive alternative to death – takes on a focused structural and affective urgency. It should surprise no one that women constitute the vast majority of shelter staff and volunteers. The role of women in the production of home, particularly in relation to domestic maintenance and emotional labor, has been extensively written about in a variety of contexts.[3] While my focus here is not primarily on gender, the status of, and kind of, social actor in this drama of home is important. These shelters are designed to produce a set of encounters between people and architecture, between people and animals, and between animals and space. These encounters, ones both intended and unintended, occur within the realm of movement, orientation, encounter, sight, smell, touch, surface and space, corridor, window. So what these spaces "say" about home and our ongoing relationships with companion animals is not expressed exclusively in words. In fact, the nonhuman inhabitants of these spaces don't speak with words at all.

As a filmmaker and visual studies scholar by training, my interest in the design, discourses, and experience of animal shelters began when I started photographing the interior of the San Francisco Society for the Prevention of Cruelty to Animals (SF SPCA) Maddie's Pet Adoption Center. Built in 1998 through funds largely provided by Dave Duffield – the philanthropic force behind the multi-million dollar private animal welfare foundation Maddie's Fund – the construction of Maddie's Adoption Center was the culmination of the SF SPCA's long institutional history and a significant index of the companion animal's

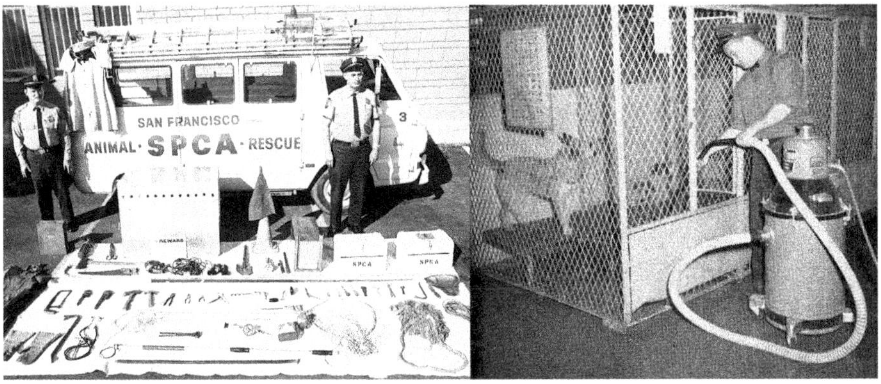

Figure 8.2 SF SPCA shelter aesthetics, pre-1980s

Source: The San Francisco SPCA archives, Special Collections, SF Public Library.

emerging status and role in American cultural life. The result of an extensive capital campaign, Maddie's Adoption Center was billed as a "new concept in sheltering." Designed to look like "trendy lofts and upscale residences," mimicking the facades and architectural vernacular of San Francisco's iconic neighborhoods, this shelter materialized an aesthetic of optimistic upper middle class domesticity. A spacious entryway opens up onto a brightly lit community room where visitors and staff mingle. Plants and sofas create an atmosphere of repose and relaxation. An overarching feeling of comfort and design is evident in the coordination of colors, the balance of sightlines, and the careful attention to detail. Just down Lassie Lane, dogs are viewable through large display windows, which look into their rooms. On the other side of the atrium, "cat condos" feature adoptable cats in communal or private rooms, equipped with toys and climbing structures and all of this viewable through the large windows that face out on to the atrium and its adjoining corridors. Sound dampening materials mute the rancor of barking, and carefully-engineered airflows whisk away any streams of unpleasant odor – acoustic and olfactory architectural features that calm the human visitors and animal residents alike.

The Aesthetics of Animal Control

Maddie's Adoption Center was the vision of its president, Richard Avanzino, who served in the executive role from 1976 to 1999, and this building is tied intimately to his interpretation of the agency's mission to "advance the human-animal bond." When Avanzino took over leadership of this private non-profit agency, the SF SPCA was still under contract to the city of San Francisco, serving as the animal control branch of government and overseeing enforcement and regulation of local laws. The agency had, in fact, acted as the animal enforcement wing of local government for most of its then 108-year history. In 1888, in response to the brutal conditions of working horses and other city animals, the Society took primary responsibility for the oversight and administration of laws relating to animal life, welfare, and death. Under Avanzino's management, the agency continued its governmental duties as officers in uniform investigated animal cruelty cases, accepted surrendered animals, rounded up stray animals, facilitated adoptions and carried out euthanasia. For the most part, it euthanized animals who were deemed unfit for adoption or, much more commonly, animals who weren't adopted quickly enough and were killed to make kennel space for the high number of animals waiting to replace them (and most often meeting the same fate). But then in 1989, after years of negotiation with the city, Avanzino relinquished the city contract in order to pursue "a new model of sheltering," and law enforcement responsibility was handed over to a newly created city agency, the descriptively named San Francisco Animal Care and Control (SF ACC).

Without the financial, administrative, and ideological burden of its animal policing obligations, the SF SPCA was freed to re-imagine itself as something else, which it did through the implementation of new and expanded "life saving" welfare programs – Animal Assisted Therapy, Hearing Dog Training, Behavior Classes, Grooming Services, Foster Programs, low income assistance, and a revitalized adoption program. These programs emphasized the bonds between people and animals in roles related to familial kinships – guardian, companion, care giver, dependent. These programs forcefully recast dogs, in particular, not as a threat or public hazard, but rather as the anchoring affective center of family life. Avanzino articulated his vision for the status of companion animals within the language of a moral imperative, re-branding the shelter itself as "no kill." His decision was a powerful enunciation of the zeitgeist in which questions of companion animal welfare were becoming exclusively yoked to questions of killing: not how to kill (as the discussion had been framed previously – decompression chamber, gas or lethal injection), but rather who or what is killable. In a widely read and influential essay published for the shelter community in 1989, Ed Duvin critiqued the conditions of the US shelter system. He wrote:

> Shelters will continue to be nothing more than processing plants until they begin the transition from sanitation dumps for the public's unwanted "baggage" to vital community education centers. Shelters cannot continue to be slaughterhouses and friends of animals cannot continue killing healthy beings in the name of mercy. A new and larger vision is needed, a vision in which shelters hold themselves accountable for meeting demanding performance standards that preserve life – not destroy it.
>
> (Duvin 1989, 6)

Duvin's vision for the future of companion animal welfare was articulated via a metonymic strategy. The rhetorical collapse of the "the slaughterhouse" (a physical location) with its meaning (culturally sanctioned and institutionally enabled killing) was deployed as a way to produce a taxonomy, a way to classify companion animals away from food animals, to linguistically render one kind of creature as killable and the other as emphatically not. Avanzino's own statement about the status of companion animals and his vision of the future of animal welfare was articulated in similarly metonymic terms but with a materialist emphasis. For Avanzino, his agency's philosophy about the value of animal life required a new building because he understood, at some level, that perceptions about the status of animals arises within the unfolding drama of bodies inside of situations, both material and discursive. The "old shelter," with its bars, concrete floors and rows of fencing, loud sounds and smells, was no longer a material or discursive possibility. Cages produced an institutionalized mode of subjectivity wherein animal bodies are subject to control and rendered visible only through the explicit machinations of power. Cages, in this context and importantly elsewhere, were the tools of docility and dependency, physical enclosures which produced a discrete inside and outside, with the control of movement, life and death for those on the inside securely dictated from without. Bars and cages are the aesthetics of incarceration, what Foucault has described as the "art of government" – the styles and practices of governmentality – and governmentality was what the SF SPCA was hoping to narrate as its historical past tense. Avanzino and his agency mulled over the question: what should this new shelter look like, be like, feel like? If the "old shelter" produced these conditions of "killability," how does one build a space in which killing becomes unthinkable? What materializes, literally, when the slaughterhouse is reimagined as something else? They needed only to look back to the future.

Figure 8.3 SF SPCAs partnership with Gump's Department Store window displays, circa 1980s

Source: The San Francisco SPCA archives, Special Collections, SF Public Library.

The Aesthetics of Desire

George Miers is a Bay Area-based architect whose résumé features a long list of public building projects including numerous North American animal shelter facilities.[4] In a 2013 interview, he recounted how he started his career in animal shelter design. In the late 1980s he took his young daughter to the Oakland Zoo and was quite horrified by the conditions under which the animals were living. As he had some ambition to work in animal housing, he sent a letter to the president of the zoo, itemizing his critiques of the animal enclosures and offering his pro bono services. Quickly and to his surprise, Miers became involved in the design and construction of Oakland's new orangutan habitat. Working in collaboration with Jane Goodall, a consultant on the project, Miers went about the business of designing a habitat that could meet the material and psychological needs of two sets of Homininae – orangutans and humans. His dilemma: how to provide enough space and mental stimulation for the orangutans while also designing for the spectatorial demands of zoo visitors. Mirroring Miers' own negative impression of the habitats, zoo visitors were increasingly unhappy with the sight of small enclosures and the prominence of bars and cages. They wanted unobstructed views but also they wanted some solace that the animals were "happy" in their spaces. Providing solace to zoo visitors took the form of climbing areas, abundant flora, and other props designed to visually reference the animals' "natural" habitat and to disavow, visually at least, their confinement. The design problem posed by the orangutan habitats was one that would come to dominate Miers' career in animal housing design: how to balance the needs of production (stimulated, active and visible animals) with the requirements of consumption (to satisfy the spectatorial and affective needs of visitors, viewers, or in the case of the animal shelter – potential adopters).

In 1990, Miers received a contract to design and build the Oakland SPCA. That shelter (now called the East Bay SPCA), built in 1992, was a pioneer structure in contemporary shelter design. The facility gestures towards what would become the defining architectural feature of the "new concept in sheltering" – large public spaces oriented around display window habitats. In the PowerPoint package that Miers presents to his potential shelter clients, he ties his design concept of this hallmark East Bay facility to his own nostalgic adolescent memory. Miers grew up in a South San Francisco public housing project and he recounts an enduring experience, a family outing to Gump's department store. At its downtown San Francisco location at 250 Post Street, amidst the hustle and flow of shoppers, adoptable dogs and cats were displayed in its storefront windows, surrounded by sumptuously fabricated scenes and invested with the exuberant energies of capital. What was this indelible memory that had such an effect on Miers' future animal shelter designs?

In the late 1970s, 30 years before they would build Maddie's Adoption Center, the SF SPCA, like every animal shelter in the US, euthanized close to or over 90 percent of the animals in its care. In an early attempt to lower that number, the agency took the animals out of their cages and began displaying them in mobile crates in San Francisco's financial district. Away from the cages and concrete floors, separated from the noises and smells of the shelter, adoption rates skyrocketed and euthanasia rates dropped. The strategy was so successful that eventually the SF SPCA partnered with Gump's Department Store to regularly display animals in their storefront windows. The exuberant scene was described this way:

> People of every size, age, lifestyle and language were united by the animals. Parents scooted their children to the front to have a look, then had to be dragged away themselves. Frequently animal fans filled the sidewalk completely, spilling other pedestrians in to the street. Barricades and cones had to be set up to control the crowds. … This time, working closely with the SF SPCA, Gump's staff had outdone themselves. They created scenes that were charming and attractive to behold, and they also designed the rooms to be luxurious accommodations for the animals. Whether in a cozy Victorian parlor, a doll house, or a forest of fanciful trees, the animals made themselves at home in their custom built environments.[5]

The window dressers and design specialists at Gump's knew something about generating excitement and desire, "hiring a good displayman is like going to a charm school. The store enters discouraged, unattractive and dissatisfied. After a few short months of display treatments it emerges streamlined, radiant – and ready for love" (Gaba 1952, 13). "Ready for love" was the narrative that the agency wanted for itself and its animals, a narrative that was open to the possibility of a happy future, one no longer over-determined by death. The SF SPCA's partnership with Gump's extended into the early 1990s and the tradition continues today in the form of the Macy's Holiday Windows, when the SF SPCA partners with that large department store to showcase adoptable animals during the Christmas buying season.[6] The tradition continues in other ways too: Miers incorporated this personal memory and significant institutional lesson in the power of commodified display into his 1991 Oakland design, integrating the architectural vernacular of the department store window directly back into the structure of the shelter, and directly into the discourses of home that circulate there. So the question, "how does one build a space in which killing becomes unthinkable?" was answered quite simply: The display window. This architectural situation – the commoditized animal displayed behind glass, surrounded by domestic props – forms the foundation of almost every new shelter designed in the United States and is replacing the cage as the defining aesthetic feature of contemporary animal welfare. Miers' Oakland shelter was a pioneering structure, but the SF SPCA, designed by the architectural firm ARQ Architects (Lucinda Schlaffer and Paul Bonacci) more fully developed and centralized the concept of the window display and emphatically articulated it as a defining architectural feature of contemporary shelter architecture. This history lesson in shelter design is also a lesson in contemporary capitalism's magical powers. It speaks to the profound effect of a simple architectural situation: commoditized display. Somewhere in the space between the homeless animal, the surface of the window display, and the gaze of an onlooker, a mysterious feeling is created. It's a feeling of captivation or objectification, perhaps. Whatever we name it, the situation produces a feeling of desire, of want – I want that dog. This positive feeling is different from the negative feelings that shelter visitors experience when confronted by an animal in a cage. In her book on the relationship between style and commerce, Virginia Postrel writes:

> Aesthetics is the way we communicate through the senses. It is the art of creating reactions without words, through the look and feel of people, places and things. Hence, aesthetics differs from entertainment that requires cognitive engagement with narrative, word play, or complex intellectual allusion. … Aesthetics may complement storytelling, but it is not itself narrative. Aesthetics shows rather than tells, delights rather than instructs. The effects are immediate, perceptual, and emotional.
>
> (Postrel 2003, 6)

The image of the dog in the window demonstrably produces an immediate, perceptual, and emotional response. Its power to produce, in this case positive, feelings is tangibly real and matched only by its power to make the narrative that brought the animal to the situation of the display window in the first place disappear. The homeless dog's appearance in the window display allows it a kind of rebirth as an object of desire, now disassociated from the social problem of pet over-population and its complicated subplots of reproduction, species, breeding, laws, responsibility, geography, culture, urbanity, poverty, abandonment, and death. Seen through the display window, the individual animal is bestowed individuality, transformed from secondhand to retail.

The (False) Simplicity of Cute

Miers narrates the contemporary animal shelter's design history from an aesthetics of incarceration to one of commodification as a teleology, the inevitable forward march of progress. In his words, the architectural shift from bars and cages to "cat condos" and "dog habitats," is the simple outcome of an "enlightened view" about companion animals. For him, these new shelters announce that the shelter dog is worthy of membership of our families, just like any other companion animal we might buy from a breeder, for example, and that we shouldn't kill them. And certainly in the case of these new shelters, euthanasia rates have been drastically lowered and adoptions have been greatly increased – the result of, I've been arguing here, a complex mix of ideology, aesthetics, and policy.[7] The historical narrative that Miers relates is, in reality, more spatial and less temporal – these new shelters are, for the most part, built in areas of relative affluence by agencies with means, while older shelters continue to operate right across the road or across county lines. The life-saving lessons of commodification have not gone unnoticed, however, as many of the older shelters – still operating under an aesthetic regime of bars, cages, and concrete floors – have partnered with local

Figure 8.4 "Cute" images of children and animals

Source: Swatt Miers Architects. Photograph by George Miers.

pet stores to display their adoptable animals inside of retail space. And so the old and the new co-exist side by side. That which lingers in the past, which exists to the side, may still call to us with its presence, registered as a faint but tangible co-perception. Which leads to the question: does the specter of an earlier aesthetic regimen and its corresponding set of ideologies and affects haunt the exuberant situation of the window display? What becomes of killability within the realm of a commoditized view of companion animals?

When looking through the visual materials that these shelters generate, the art on the walls and the images on their websites, children become remarkably omnipresent. At first impression, one registers how children serve as icons for the notion of family and home that circulates in these spaces. But it's not only the imagery of children, or children with puppies, that manifests a kind of omnipresent neotony – the spaces themselves are painted in bright colors, and there is an overall soft-edged and pastel hue in the fabrics and furniture. In many of these shelters, cartoon-like statues of dogs and cats line the walkways and adorn the lobby. Scale, the relation of objects to size and space, is at play in the design of many of the architectural features. The omnipresence of these types of features invests the scene with an overbearing aesthetic of what can only be described as "cute." The cultural critic Sianne Ngai has recently defined "cute" as a ubiquitous aesthetic style that indexes a situation of suspended agency. She writes:

> Whether in response to socks or to large scale, mass mediated spectacles of public intimacy, cuteness solicits a regard of the commodity as an anthropomorphic being less powerful than the aesthetic subject, appealing specifically to us for protection and care. As Merish puts it, the cute "always in some sense designates a commodity in search of its mother" thus "grafting commodity desire onto a middle class structure of familial, expressly maternal emotion."
>
> (Ngai 2012, 60)

The aesthetic repertoire of cute fulfills the welfare goal of configuring dogs and cats as in need of care and protection and yoking them to narratives of childhood innocence. But cuteness exudes a false simplicity. Layered inside of its soft associations and adorable features are other more hard-edged and even menacing narratives. Ngai argues that cuteness is, "a commodity aesthetic, with close ties to the pleasures of domesticity and easy consumption. As Walter Benjamin put it: 'if the soul of the commodity which Marx occasionally mentions in jest existed, it would be the most empathic ever encountered in the realm of souls, for it would have to see in everyone the buyer in whose hand and house it wants to nestle'" (Ngai 2011). The dog in the display window is, perhaps, the commodity form par excellence, one that we wish to hold in our hands and nestle into our houses. This status of this creature is apparently the opposite of killability. The dog and our feelings towards her are a refined form of domestication wherein the animal serves as the lively and fleshy agent of a naturalized commodity fetishism. The dog's presence in our lives fulfills the desire to simplify our relationships to commodities even as it situates the commodity deeper in to our hearts and homes. For Ngai, though, the cute also indexes ambivalent, explicitly contradictory, or even aggressive feelings. She encourages us to see cuteness as a way of aestheticizing powerlessness as it hinges on a sentimental attitude toward the diminutive or weak, which is why cute objects – formally simple or non-complex, and deeply associated with the infantile, the feminine, and the unthreatening – get even cuter when perceived as injured or in need. So there's a sadistic side to this tender emotion, one directed at the shelter dog in the display window. Cuteness is, perhaps, a way to kill with kindness. Although certainly these two aesthetic extremes – cages and display windows – are not equivalent forms of violence they may be closer in effect than my pithy cliché about killing with kindness suggests. In the case of the SF SPCA, the agency does not accept animals that they deem "unadoptable," which very often includes breeds such as pit bulls and mastiffs, animals that uneasily assimilate into the aesthetic imperatives of cute which these habitats perform or even demand.[8] These dogs, physically large and associated in the popular imagination with violence, dog fighting, and urban masculinities, are not easily interpolated into the décor and its narratives and their exclusion most often produces a deadly result. These dogs, even though they are disproportionately represented in the shelter system generally, are not proportionally made available for the particular version of *home* that these shelters perform.

A Hole in the Glass

As I'm interested in investigating my subject of animal housing through the specific material effects and affects, the brick and mortar consequence of our lives with dogs, I ask you to direct your eyes to the images in Figure 8.5. As I've argued throughout this paper, the shelter animal, within the regime of new shelter aesthetics, has become the object of a commoditized and affective gaze. But that's not the whole story.

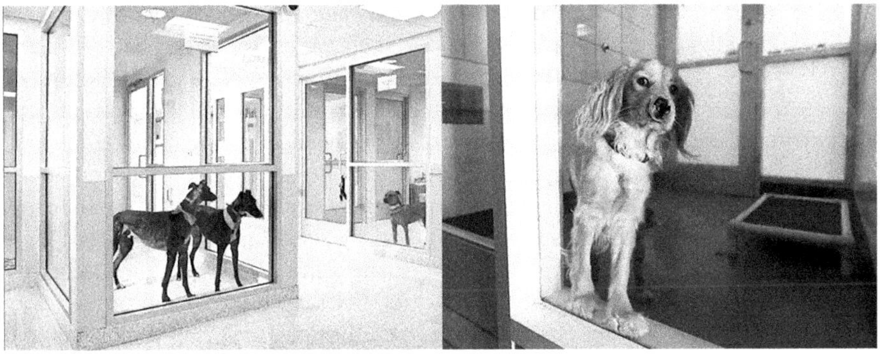

Figure 8.5 Large glass windows emphasize the visual dimension of animal display

Source: © ARQ Architects – Architect.

The images in Figure 8.5 embody a palpable tension present in these spaces: the dog, on the one hand, as an object of our gaze and, on the other, as an agent of looking. As Derrida and his cat have so persuasively argued, an animal can address us: “The animal looks at us and we are naked before it. Thinking, perhaps, begins there” (Derrida 2008, 15). This is how Derrida begins his analysis into how the threshold, the limit, between the human and the animal has come into being. He wants to know how the plurality and immensity of nonhuman life came to be contained within one single designation – the animal – a word which has effectively sequestered the human within an esteemed space of reason and language and has violently relegated the animal to a space of unthinking: docile without language. Derrida’s intervention into this situation is in the form of what one could call *a reality test*. An encounter with a real cat. Not a metaphorical animal, he tells us at length, but a real small cat. Derrida’s encounter and the knowledge it produces is one between two agentic creatures, conjoined in a process of exchanged glances across the threshold of language and ontology. In her analysis of multispecies histories and our ongoing lives with animals, Haraway summons Derrida’s encounter with his cat. For Haraway, Derrida’s insight is significant but incomplete. She writes, “But with his cat, Derrida failed a simple obligation of companion species; he did not become curious about what the cat might actually be doing, feeling, thinking, or perhaps making available to him in looking back at him that morning” (Haraway 2008, 20). So, what do dogs in the shelter, living in their display habitats and laboring under the sign of commodity, feel or think?

As an attempted answer to this question, I direct you to a small but significant architectural feature that suggests how shelter animals respond to the situation that they find themselves in. I direct your eyes back to the two images of dogs on display (Figure 8.5). The left-hand image announces the emphatically visual orientation of these spaces – dogs in display windows gazing at each other. Or more accurately, dogs posed in windows, exchanging glances with each other while being made totally available for our scopophilic capture. The right-hand image affords us a more direct view and we can now see that near the bottom of the window (in front of the dog's nose), a small hole has been drilled into the display glass, encouraging us to hold our fingers here, allowing this dog to sniff; encouraging fingers to be touched by tongues, to facilitate a tactile encounter of aroma and touch. In a place where encounters between people and dogs are engineered to occur as primarily spectatorial, this is a remarkable detail in that it's a concession to the fact that dogs are not primarily motivated by sight; rather by smell. It is an acknowledgment to the fact that every encounter is a 2-way exchange (or more, 3-way, 4-way) between entities encountering each other and doing so with their own sensorial and cognitive capacities and from their own physical, political or cultural position. This hole in the glass, a small architectural detail, reveals the ways that these shelters, simultaneous with their emphasis on the commoditized and scopophilic consumption of animals, are also designed with active and sensing dogs in mind. For the most part, the explicitly animal-centric spaces are ones that the public never sees: darkened and quiet rooms where the stressed and fragile animals are first brought in to the shelter, exercise

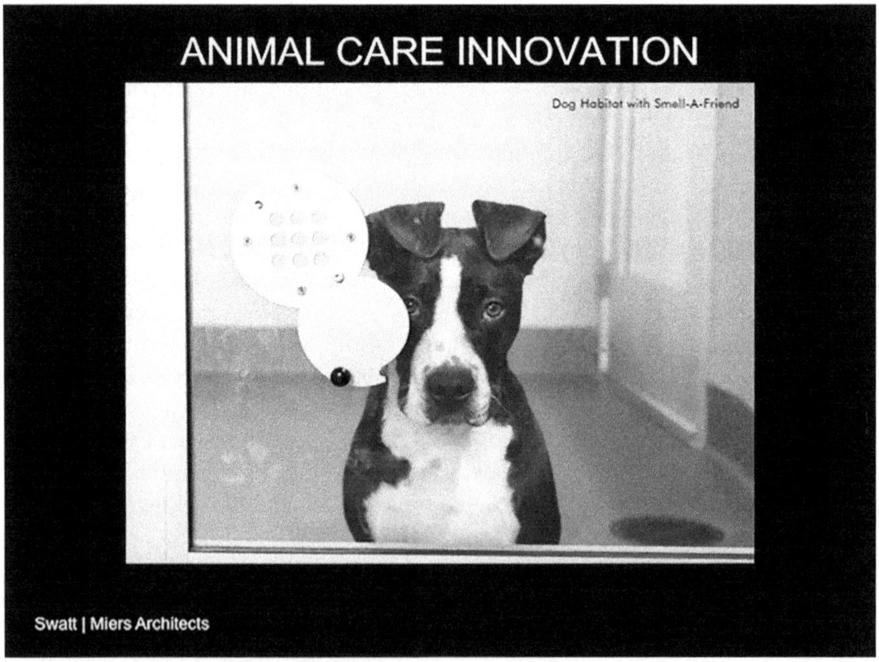

Figure 8.6 Smell-A-Friend from George Miers' Powerpoint Presentation "Animal Sheltering for the 21st Century"

Source: Swatt Miers Architects. Photograph by George Miers.

facilities, training facilities, for example. But the hole in the glass is an important feature and perhaps an important interruption into the semiotic and affective logics of the commodity that circulate in the public areas of the shelter, that constitute the structure and experience of the display habitat. The hole in the glass affords the dog a chance to respond, in the form of a sniff, a lick, or even a bite. And every one of these unpredictable responses is enacted with ordinary regularity, every day. Most often, however, these holes get closed over when fingers are bitten or when people with malice towards the dogs pass poisoned or dangerous objects across the threshold of the glass.

In response to the unscripted and possibly dangerous encounters that are possible in this small opening in the threshold, George Miers has developed a trademarked solution: the Smell-A-Friend (see Figure 8.6). In yet another circularity, an earlier architectural feature finds its way back into the present. In this case, the bars and barriers of the "older" shelter reappear, closing over the hole in the glass and constraining the contact between the creatures on each side of the glass. It's curious that the Smell-A-Friend is shaped, consciously or unconsciously, like a speech bubble. What does the dog have to say? Whatever the dog has to say, it will be communicated from inside the aesthetic situation of simultaneous commodification and confinement. And it will not be spoken with words. As Haraway writes, "The truth or honesty of nonlinguistic embodied communication depends on looking back and greeting significant others, again and again. This sort of truth or honesty is not some trope-f—free, fantastic kind of natural authenticity that only animals can have while humans are defined by the happy fault of lying denotatively and knowing it" (Haraway 2008, 27). If we become curious about what the dog has to say, as Haraway argues we are obligated to do, we should focus our attentions on the hole in the glass. It allows us to see, as if we are watching ourselves in a movie, how we are situated in relation to the creature in front of us. Our experience of this space, these surfaces, this situation, and this animal allows us to see and feel this encounter in all of its complexity: an image, an instance of domestication, a soft wet tongue or a piercing sharp tooth, a creature that is subject to the collective histories of companion animals, a creature that is subject to its own personal history, here is the cute commodity we long to hold and bring in to our homes, here is the social problem that we want to disappear, and here is the creature who looks back. This last point is crucially important if we wish to engage in critical practices that more fully consider species, intersecting with class, gender, race, and sexuality, as central to the making of worlds. We should focus our attention on this hole in the glass, as thinking with animals occurs there.

Notes

1 The San Francisco bay area includes numerous cities and towns and there are many animal shelters, built within the past 15 years: Peninsula Humane Society in Burlingame (Miers, architect), Humane Society of Silicon Valley in Milpitas (Miers, architect), Tri Valley SPCA in Dublin (Miers, architect), East Bay SPCA in Oakland

(Miers, architect), Animal Rescue Foundation (Miers, architect), The San Francisco SPCA in San Francisco (ARQ, architect), Berkeley Animal Care Services (Burk Toma Architects, architect). But there are also numerous older shelters, dating from the 1940s to the 1970s, located in municipalities that lack the resources, political will or clout, to build a new shelter: Oakland Animal Services, San Francisco Animal Care and Control, for example. Many of the older shelters are city-run agencies that do not have access to private or corporate donor networks. My analysis does not exhaustively account for each and every building; rather, it speaks to them in aggregate. My analysis is anchored most specifically in the history and architecture of the San Francisco Society for the Protection and Care of Animals (SF SPCA), a pioneering structure and agency.

2 Donna Haraway's definition of a companion animal from her lecture titled, "Birth of the Kennel," delivered at the European Graduate School in August 2000. Available from https://www.youtube.com/watch?v=-yxHIKmMI70 [Accessed 1 February 2016].

3 See Rothschild (1983); Ehrenreich (2001); Gaarder (2011).

4 A complete list of Miers' public building projects is available at his firm's website: www.swattmiers.com/#/projects/animalcare/0/. His numerous greater bay area animal shelter facilities include: Peninsula Humane Society (Burlingame), Humane Society of Silicon Valley (Milpitas), Tony LaRussa's Animal Rescue Foundation (Walnut Creek), Tri Valley SPCA (Dublin), City of Oakland Animal Shelter (Oakland), East Bay SPCA (Oakland), Contra Costa Animal Services Facility (Martinez).

5 From the San Francisco SPCA publication, *Our Animals*, Summer 1984.

6 For links to the current SF SPCA Macy's windows program, see http://sfspca.org/support/events/macys-holiday-windows.

7 I am necessarily glossing over the myriad details and specificity of shelter policies that, in conjunction with shelter aesthetics, play a determining role in "live release rates" – or the percentage of animals euthanized at the shelter. One significant factor is if the shelter is "an open door" or a "selective admission" facility. "Selective admission" shelters only accept dogs who have a high probability of being adopted. The SF SPCA is a selective admission shelter and became so at the time when it relinquished the city contract and became a "no-kill" facility. The SF ACC, located right next door, is an open admission facility and accepts every animal that is surrendered or enters via the enforcement wing of the agency.

8 The number and breed of animals accepted and declined by the SF SPCA is debatable, as accurate and long-term numbers are not kept. My assertion is based on anecdotal evidence (kind and number of dogs available on the website) and on interviews and discussions conducted with San Francisco ACC shelter workers.

References

Ahmed, S. (2006) *Queer Phenomenology*. Durham, NC: Duke University Press.

Cassidy, R. (2007) "Introduction: Domestication Reconsidered," in R. Cassidy and M. Mullin (eds), *Where The Wild Things Are Now: Domestication Reconsidered*. Oxford: Berg, pp. 1–29.

Derrida, J. (2008) *The Animal That Therefore I Am*. M.-L. Mallet (ed), D. Wills (trans). New York: Fordham University Press.

Duvin, E. (1989) "In the Name of Mercy," originally published in the journal *Animalines*. Available online at: http://doggymomblog.files.wordpress.com/2012/05/ed-duvin-article-1989.pdf.

Ehrenreich, B. (2001) *Nickel and Dimed: On (Not) Getting By in America*. New York: Metropolitan Books.

Gaarder, E. (2011) *Women and The Animal Rights Movement*. Piscataway, NJ: Rutgers University Press.

Gaba, L. (1952) *The Art of Window Display*. New York: Thomas Crowell.

Haraway, D.J. (2008) *When Species Meet*. Minneapolis: University of Minnesota Press.

Mulvey, L. (2001) Criterion Collection liner notes for *Written On The Wind*, directed by Douglas Sirk. New York: Criterion Collection, DVD.

Ngai, S. (2011) "Our Aesthetic Categories: An Interview with Sianne Ngai," *Cabinet*, 43. www.cabinetmagazine.org/issues/43/jasper_ngai.php.

Ngai, S. (2012) *Our Aesthetic Categories: Zany, Cute, Interesting*. Cambridge, MA: Harvard University Press.

Postrel, V. (2003) *The Substance of Style: How the Rise of Aesthetic Value is Remaking Commerce, Culture, and Consciousness*. New York: Harper Collins.

Rothschild, A. (1983) *The Managed Heart: Commercialization of Human Feeling*. Berkeley: University of California Press.

9 Concrete Kingdoms

Heini Hediger's Territories at the Zurich Zoo

Christina Katharina May

In 1942, Heini Hediger published *Wildtiere in Gefangenschaft* [Wild Animals in Captivity],[1] one of the first scientific handbooks to engage with keeping nondomestic animals in zoos. Perhaps his best-known and most provocative statement, frequently cited in biological and cultural studies of zoos, was the relativism of captivity.[2] Each animal is a prisoner of its surroundings, Hediger wrote, a prisoner in space and time. Groups and individuals are enclosed by invisible but powerful barriers of their territory; they are confined by geographical distributions and by their relationship with other animals (Hediger 1942, 15; Hediger 1964, 5). Freedom, he stated, was a rather anthropocentric concept: almost all varieties of animals (in his various publications he referred to crustaceans, birds, or mammals) are content when they live undisturbed in a certain area, their territory. As a consequence, the size of an area – the amount of space – was not relevant, but rather that the quality of its environment guaranteed welfare. Zoo animals, similar to animals in the "real wilderness," were not prisoners but landowners. Territories are marked by the movement and the traces of bodies; in addition, the body itself is enclosed by invisible barriers, which Hediger introduced as "flight distance." Flight distance is the minimal distance between an individual and its enemy; as for zoos, between the zoo animal and the visitor, keeper, or other animals. The abstract concept of natural environments allowed various interpretations for its implementation and design. Hence, the development of zoo buildings provided the chance to reconsider planning processes, and to rethink aesthetics as regards the psychological and physiological needs of its users. But how does the transformation of territories defined by movement and behavior work with regard to zoo design? Did Hediger implement "zoomorphic" instead of anthropomorphic zoo architecture?

Hediger played a central role in defining the tasks of zoological gardens and reconsidering their operation and construction processes after the Second World War. Participating in major international meetings of zoo directors, he was among the founding members of the International Union of Directors of Zoological Gardens established in Rotterdam in 1946. Hediger served as its secretary and ultimately led the Union in 1949. He started his career as a zoo director in Bern at the Tierpark Dälhölzli. From 1944, he headed the Zoological Garden of Basel and prepared a layout for its new general plan before moving to Zurich in 1953 to supervise the redevelopment project of the city's young and underdeveloped zoological garden (Rübel 2009, 20–29).

Hediger established a new scientific field named "zoobiology" in which he propagated principles of animal husbandry, an early state of behavioral enrichment practices, and, the main topic of this article, *a concept of natural space*. His notion of space relied on the development of the new biological field of animal psychology, as well as on observations of animal behavior in the zoo and in the field, mainly in African regions. Basing these principles on studies by his ethologist colleagues, mainly Nikolaas Tinbergen and partly Konrad Lorenz, Hediger introduced new standards for keeping zoo animals according to ethological studies and observations of free-roaming species and captive individuals. In addition, he wished to address zoo visitors, and tried to develop exhibits that fostered educational messages about behavior and social life.

Hediger's topological interpretation of space was widely accepted, widely interpreted, and, willingly or not, even widely misinterpreted because, while publishing numerous texts, he rarely commented on concrete design. Stating that the size of a cage would not matter as long as the animal was tamed and has lost its flight distance, Hediger's statements legitimated hygienic, but small and bare exhibits of the zoos' so-called "disinfectant period" of post-war modernism, especially after the English translation and the American reception of his book in 1950 (Hyson 1999, 443–446). The topological abstraction of natural space offered multifaceted possibilities to transfer it into built space. Even Hediger's soundbites "Nature means more than the sum of an infinite number of containers of space (cages), however natural" (1942, 76; 1964, 72) or "The cube, indeed a straight line of any kind, is unbiological" (1970, 21) led to misinterpretations. While the latter phrase hints at architectural forms and at a range of contemporary architectural movements, Hediger carefully avoided suggesting design in terms of aesthetics. Nevertheless, during the 1960s various of his texts recommended "natural" features to a greater extent because Hediger became interested in the appreciation of the visitors. His advice seems to change along with his growing experience in planning, designing, and managing zoos.

Thus, it is vital to take a closer look not only at Hediger's texts, but at the buildings and landscapes he was responsible for during his directorship at the Zoological Garden of Zurich from 1953 until 1973.

The purpose of this study is not to show that Hediger failed to realize his documented principles in the physical world, especially in his constructions, but to reveal the relationship between Hediger's topological principles and their spatial and visual implementation. The questions to be asked, however, are how Hediger's concepts of space were translated and transformed into the three-dimensional building space. Aspects such as the contemporary architectural style and developments of the human–animal relationship influenced the final results. By introducing the buildings of the Zurich Zoo of the 1950s and 1960s, the study examines a key period of the history of zoological gardens, which is still largely undiscovered because it is discarded as a disinfectant period.[3]

The most important building of this period seems to be the Africa House at the

Zurich Zoo, which opened in 1965 and housed rhinoceroses, hippopotamuses and several species of birds. In his crucial publication, *Mensch und Tier im Zoo* [Man and Animal in the Zoo] (1965), Hediger refers to the Africa House as a role model. He stated that the Africa House would incorporate all aspects of his theoretical approach to designing artificial environments (1965, 28). The building combines two different interpretations of nature: for visitors, it represents visual features; for animals it offers topological qualities of natural space. The Africa House comprises specific aspects which are important for zoo design in general, like the disposition of sight lines, the composition of boundaries between visitors and animals, and the interpretation of animal space and natural conditions. Assuming that the building marks a shift from the international style buildings of post-war modernism towards built structures that integrated traces of environmental consciousness, the article at first takes a closer look at the planning at the Zurich Zoo of the 1950s, which reflects the adaption of Swiss post-war modernism. This leads to the design and building of the Africa House, and ultimately, the Otter Enclosure of 1970 which serves as a subsequent example for a rather naturalistic interpretation of natural space. The text concludes with a brief look at Hediger's adaption in terms of human architecture and ideas about social and physical spaces.

The Nature of Zoos

Observing living animals in a somewhat natural context was intended to educate the public not only intellectually but also to boost collective morale. Interpreting naturalistic landscapes as pleasant and educational spaces influenced the design of zoos as landscape gardens. These aesthetic demands had already accompanied the planning of the first public and scientific menagerie, the Jardin des Plantes in Paris, established in 1793, and it accompanied the evolution of zoos in the nineteenth century (Baratay and Hardouin-Fugier 2004, 73–77). Hagenbeck's Tierpark established a new dimension of grand-scale illusionistic sceneries of exotic regions at the turn of the nineteenth century. Zoologists appreciated Hagenbeck's implementation of moats, which provided an unobtrusive and stage-like view, although many of them discarded the naturalistic scenery as charlatanry.[4] While Hagenbeck's "revolution" seems to have brought the illusion of nature into the zoo, Hediger searched for a less anthropomorphic, respectively mimetic interpretation of nature, looking rather for a systematic version. Thus, it is surprising that Hediger's role in zoo history is mainly one as an advocate for providing natural conditions as opposed to technical and hygienic buildings with tiles and concrete. Still, nature is often identified with plants and somewhat irregular structures, which definitely depart from Hediger's early intentions of the 1950s. Nevertheless, the zoo's nature has always been intended to please the paying public, whose ideas about nature often differed from zoological interpretations and educational messages. Hediger always claimed, however, that the zoo is a cultural institution, and should be appreciated like a museum or theater. Thus, the zoo gained wide appreciation because it was the place that presented and represented nature in the city, or, to use Hediger's phrase: "The zoo is the emergency exit to nature" (1965, 80).

Therefore, the representation and construction of nature in zoological gardens depends on socio-historical concepts, ideas about ecological order, and imperialism (Kohlstedt 1996). Analyzing built structures of zoological gardens is vital in order to understand the institutions and their inherent constructions of human–animal relationships. Mullan and Marvin argue that "housing" was an anthropocentric term, and recommend studying the "relationship between man and animals in terms of the physical structure in which animals are housed in captivity" (Mullan and Marvin 1999, 46). Gruffudd has shown the complex connection between London's constructivist animal exhibits, contemporary architectural modernism, and ecological concepts, which Tecton's Penguin Pool echoed. The housing conditions not only reflected kinetic interpretations of organisms, but also concepts of human housing. The London Zoo and Dudley Zoo became places to implement innovative design, biological, and even sociological experiments (Gruffudd 2000, 234). This phenomenon is also true for the buildings of the London Zoo in the 1960s, when Cedric Price developed structures such as the Snowdon Aviary. Steiner relates the experiment of engineering to Price's activity with Archigram, and again the zoo serves as a very special location (Steiner 2003, 8). Nevertheless, neither architect seemed to care much about empirical studies on animal behavior and physiology, but rather about theoretical concepts of organisms and ecology. In recent years, scholars have pointed out that the practices of husbandry and exhibits in zoos have to be analyzed to gain new insights into the issue of captive wild animals, which differ from interpretations at a symbolic or stylistic level. Moreover, the exhibits do not only represent certain characteristics of a species or its original habitat, but the structures and their usage manipulate animal behavior, and present images of individual animals and their relation to the visitors at multifaceted visual, topological and performative levels (Burt 2002; Uddin 2015).

Territorial Spaces

Zoologists have always claimed that the presence of living animals was the major difference between zoos and natural history museums (Rothfels 2002, 17). At the same time, zoo animals did not necessarily show "life" but stereotypical zoo behavior: Begging behavior, how animals took shelter, slept, or did not move at all. Stereotypes spoiled the attractiveness of captive animals and affected, along with viewer aesthetics and the creatures' welfare, the reputation of the institution. Hediger tried to avoid the presentation of animals as passive objects because it lacked an educational message. He criticized zoos with gallery-like arrangements of cages and enclosures, which isolated the animals from their social and environmental context. Instead, exhibition design for living animals and educational messages were expected to respect each individual animal's needs as far as time (waking hours of activity, for example) and space (in addition to the animal's size, its average range in the wild), and its physical, social, and behavioral needs. The public was intended to learn new ways of viewing because exhibit structures that respected the space-time concepts and integrated shelters did not guarantee that the animals were permanently visible. Hence, a zoo visit should prevent the fast visual consumption of the animals on display. Hediger suggested that visitors would wait to see active animals and learn about the behavior of each species. By accepting contingency, each zoo visit would become unique, and the animals appeared to have a choice of being observed or not. Thereby, the animals would not appear as mere objects but as acting subjects – "not only the animal itself, but the animal in relation to time and space" (Hediger 1942, 182; 1964, 176).

Hence, territorial behavior became important with regard to architectural and social space. Hediger's findings were not new but were influential. He derived semiotic concepts of space from Uexküll's Umwelt theory, based on environmental perception and use (Turovski 2000, 380–381). American ornithologists of the 1920s coined the term "territory" for zoological studies (Hediger, 1942, 18; 1964, 10; Burt 1943), and Hediger adopted it in *Wildtiere in Gefangenschaft* in 1942. The year of the publication paralleled World War II, and the debates about territory might have influenced the emergence of the term during the 1930s. Hediger, however, did not try to apply territorial theories on human society in his contemporary writings.

The term "territory" implied that individuals or groups were bound to a specific site, which the inhabitants controlled. Social relations between group members or competitors determined the boundaries of the territory (Hediger 1942, 15–31; 1964, 7–26). The shape of the territory resulted from a network of movement patterns on certain tracks, which led from one point of interest to another. Connecting the lines, respectively the tracks, between the functional places created a polygon. That is why Hediger stated later that natural space was never rectangular but polygonal (Hediger 1965, 27). During the 1940s and the beginning of the 1950s, however, he did not recommend the polygon for the layout of animal enclosures but tried to improve the quality of space by variations of furbishing and animal training. The abstract concept of territory is well suited to inventing building and circulation structures that rely on contemporary concepts of human topologies. The schematic illustration of tracks originates from city planning and the development of complex infrastructures. At the beginning of the twentieth century, radial structures of transportation and circulation changed into networks, which connected dwelling, working, consumption or recreational areas. Therefore, modernist debates about function sharing, complex infrastructures, and traffic patterns became relevant for the organization of social life in artificial environments, and reflected human agglomerations and the second nature of zoos.

Planning Practices

Hediger argued that collaborations between architects and zoo biologists were vital to creating satisfactory animal houses (Hediger 1949, 22). Thereby, architectural space was based on the latest findings of animal husbandry, whereas the architect had to subordinate aesthetic considerations (Hediger 1942, 177). Appreciation of individual architects was a rare characteristic among zoo directors because they usually warned against architects (Hediger 1949, 22). However, the architects Willy Kehlstadt and Max Ernst Haefeli, who worked with Hediger in the 1940s and

1950s, planned impressive zoos, resembling small, autonomous cities, but they hardly had a chance to build them.

At first, Hediger tried to implement his theoretical considerations at the Zoological Garden of Basel, where he designed an *Idealplan* together with the rather conservative architect Kehlstadt (Rübel 2009, 27). The plan implied building over the nineteenth-century landscape park with big animal houses and wide pathways. Before any major building could be achieved, Kehlstadt died, and Hediger quit as zoo director and left for Zurich in the autumn of 1953. Here, together with Max Ernst Haefeli, he started a new planning project for the redevelopment of the zoo. In 1954, Haefeli's designs relied much more on the style of New Modernism than Kehlstadt's did, but, similar to the plan for the Basel Zoo, the new general plan for Zurich suggested huge building structures, each adapted for the exhibition of specific species.[5] Hediger travelled to the United States in 1950 and Haefeli visited American zoos in 1954. Collecting ideas for the general plan of Zurich Zoo, they were impressed by huge, clean, and arena-like buildings that guaranteed hygiene and good visibility of the animals.[6]

The implementation of this plan also failed because the costs were too high, and Hediger got into severe conflict with the zoo's board (Röthlin and Müller 2000, 137–158). Nevertheless, both general plans shared the idea that visitor circulation had to be developed and that the houses supported wide, arena-like viewing areas for the public, and hygienic boxes for the animals. Each of the zoo's new buildings was considered a species-specific biological furbishing, at least in the imagination of the zoo director.[7]

The plan for Zurich Zoo integrated some innovations from exhibitions or amusement parks: pathways were widened and arranged in a polygonal layout, which corresponded to the projected animal houses. Heliclines would bridge the steep ascent of the terrain to make walking more comfortable for the visitors. In addition, an electric train would carry them through the zoo. The idea for the train originated at the Swiss National Exhibition of 1939, and was still considered a progressive means of transportation in 1954. The circulation system with its urban heliclines has never been realized, whereas the electric train went on its first lap around the zoo in 1963 (Röthlin and Müller 2000, 167).

Although the zoo was situated outside the city, at the edge of a wood and with a view of glaciers, the plans did not intend an imitation of nature. The ideas for circulation referred to an urban traffic system to guarantee the smooth and comfortable movement of the masses. At the same time, aesthetically oriented preservationists were discussing the establishment of additional national parks in Switzerland, which had to be set up according to "complex translations" of the national and local characteristics (Kupper 2012, 81). The aesthetic of wilderness did not play any role in the zoo landscape in Zurich during the 1950s and 1960s, but rather the aesthetics of "complex translations." The zoo was explicitly a man-made, second nature (Hediger 1965, 65–80).

The only building Haefeli realized was the Ape House, which opened in 1959 after a planning period of six years. The Ape House consisted of building blocks staggering horizontally at the steep slope of the site (Figure 9.1). The structure

Figure 9.1 Entrance area of The Ape House, Zoological Garden of Zurich

Source: Beringer & Pampaluchi/Zoo Zürich, 1959.

generally followed the exhibition concept of a gallery although the visitors could view the apes and monkeys from different levels and elevations. Hence, the structure provided environmental enrichment: the apes received specially designed "ape furniture," concrete cubes, which were not expected to resemble human furniture. Skylights illuminated the cages with daylight, by which the apes seemed to be on stage and were highly visible. Glass separated the visitors from the apes, and other additional precautions supported a controlled, hygienic environment. The design of the Ape House represents progressive technical and scientific developments of modernist architecture, and demonstrates several ideas about animal management, paying special attention to their health, hygiene and optimization of movement – what Chrulew refers to as Hediger's biopolitical practices (Chrulew 2013, 226). It would be an oversimplification to be satisfied with the building's classification as a modern, functional building. Spending 1.2 million Swiss francs on the animal house provoked criticism. Even worse, the accurate building structure recalled styles of contemporary cultural buildings. This typological sacrilege annoyed at least one anonymous critic (Anonymous 1959). Finally, the Ape House obtained official appreciation by receiving an award for "good buildings" in Switzerland in 1960 (Hildebrand et al. 2007, 388), which demonstrated that the status of an animal house as a cultural building, and a complex building project became widely accepted.

The Africa House: A Building for Rhinos and Hippos[8]

In 1965, the same year as he opened the Africa House, Hediger published his seminal work *Mensch und Tier im Zoo* [Man and Animal in the Zoo], which consolidated and spread his theories about territory, flight distance, and husbandry. The free structure of the Africa House is characterized as an ideal role model because, according to Hediger, it was a unique invention of a building that respected the living animal in it. Moreover, the architect combined segments of a circle to generate a structure that suggested flexibility. The rounded forms of the Africa House illustrated Hediger's approach – that buildings for animals differed from rectangular human architecture, especially human dwelling spaces. This observation also parallels statements by the Austrian-American architect Richard Neutra, whose writings on "human habitat" and his invention of the term *biological realism* influenced Hediger's architectural visions (Neutra 1956). Nevertheless, Neutra designed residential houses for humans, which is a crucial but only implicit point in Hediger's argumentation. In truth, the Africa House is an exhibition pavilion, which demonstrates animal space to the visitors.

The new house for rhinoceroses and hippopotamuses opened on 16 July 1965 (Schmidt 1967). It is situated at the southwest of the zoo, close to the Ape House. The original plan by Haefeli intended a cubic structure with huge skylights and a terraced visitor area. In 1961, Rudolf Zürcher and the landscape architect Albert Zulauf, part of a younger generation of architects, took over responsibility for designing the zoo. They continued to use the polygons for the pathways of Haefeli's plan, but they altered the projected high-rise architecture according to contemporary criticism on modernism and its cubic monotony. The house was expected to show the animal in its specific, "natural" context by providing biological shapes, social behavior, and ecological features, e.g. symbiotic grouping with free-flying cattle egrets and elements that topologically referred to the original habitat of the species. The demonstration of symbiosis was an especially outstanding characteristic (Hancocks 1971, 135). The building was originally named the *Nashorn- und Flusspferdhaus*, or Rhinoceros and Hippopotamus House, which meant that it did not refer to an absent "Africa" but informed about the major important species on display. Most of the exhibited animals were originally imported from the African continent: the two pairs of black and white rhinoceroses, the pair of hippopotamuses, the shoebills and free-flying birds. The South American tapirs spoiled the concept of an Africa House, but they had similar needs in terms of husbandry. Today, this extraordinary structure is an architectural heritage site and generally preserved. In 2015, the rhinos are to be replaced with kangaroos because the house does not meet actual holding standards (Zoo Zürich 2015).

As Figure 9.2 shows, the main path for visitors parallels the horizontal expansion of the building structure, and secondary pathways lead up the ascending slope to the entrances. The main attractions of the house, the rhinos and hippos, are not yet visible. Instead, the material qualities are on display. Reinforced concrete dominates the main façade on the southeast of the Africa House, although shrubs

Figure 9.2 Main facade of the Africa House, Zoological Garden of Zurich

Source: Zoo Zürich, 1965.

and trees have grown in the meantime and have covered the concrete relief of the front. Arch-shaped skylights tower over the façade and introduce concave curvatures as the main topic of the built structure. The skylights are also important features to provide daylight for the plants in the interior. But if that had been the only reason for the shape of the roof, simple shed roofs would have been equally effective. Hence, the building with the sculptural qualities of its roofs clearly conveyed the message that it represented the cultural function of the zoo. The towering arches are reminiscent of buildings with a cultural purpose, such as contemporary exposition pavilions (Heinsdorff 1970, 936–938). Therefore, the Africa House not only exhibits wild animals in captivity, but it serves as the manifesto of Hediger's theoretical program. The nature of zoos is not a visual imitation of the wild, but the transposition of environmental qualities (Hediger 1942, 76; 1964, 72). This is why the curvature dominates the shape of the building. Segments of a circle are essential for the layout of the whole structure. Both topics, the exposition and the curvatures as expressions of a natural-shaped space, are continued inside the building and they also determine the outdoor enclosures.

Inside the house, the visitors' area follows the sweeping curve of the main façade. Four U-shaped boxes face the visitor area. Originally, visitors had an unobstructed overview of the boxes and the animals, which were presented in a stage-like setting. A moat separated the visitors from the rhinoceroses, and only a low wall with plant boxes enclosed the pool of the hippopotamuses. The pool was exactly the size of two grown hippo bodies. Plants referred to a "tropical" atmosphere, at least symbolically, because they only grew strictly controlled in pots and boxes. The design should contextualize the exhibited animal to convey informally information about the animal's natural behavior. Zürcher added another box for the rhinos in the backstage area soon after the opening of the house because it was only possible to separate the animals at night when the keepers shifted them into the chutes. Apart from the lack of quantitative space, the design demonstrates combinations and irritations of aesthetics and husbandry. It fostered a reflection on the conventional use of cubes for housing and on the development of a semiotic program that not only met the physical needs but also created meaning from the perspective of a rhino.

According to Hediger, the interior design transposed the "wildness" of natural landscapes (1965, 28). The "wildness" is still visible today: The surface of the floors in the animal area is modeled irregularly to provide sensuous impressions for the rhinos (Figure 9.3). It rises slightly towards the back walls of the boxes, where

Figure 9.3 Interior boxes of the Africa House, Zoological Garden of Zurich

Source: Zoo Zürich, 1965.

doors lead to the chute and the outdoor enclosure. The egg form of the doors correlates to the morphology, specifically the outline of a rhino, because wild rhinos trace these oval forms through the thicket. The soffits of the round-shaped doors however are right-angled and sharp-edged, so that the rhinoceroses still could hurt their skin while passing the doors. Early photographs of the interior boxes show that the keepers have rasped off the lower part of the door edges. This example shows that daily routines of husbandry revealed the failure of the concrete form, and its improvement did not result from scientific observations and considerations but from practical handling of the animals. Nevertheless, the keepers worked mainly in the backstage area behind the curved partition wall, and were only visible to the public when they cleaned the boxes. Practices of husbandry were not part of the scenes on display. The actual effects of the round spaces on the behavior and health of the animals might be doubtful. But Hediger's interest in semiotic aspects of space might lead to the interpretation of these doors as the familiar shape of the routes of the rhinos (Hediger 1965, 28). In contrast to the direct visual contact, a ventilation system aspirated the odor of the animals. Although the ventilation has never worked sufficiently to prevent the smell of feces in the visitor area, this olfactory separation shows that not every aspect of keeping animals was intended to receive the visitors' attention and contribute to learning about animals. Still, hygiene was an important feature of veterinary medicine but in this special case, the clean environment was a crucial part of the exhibition Hediger learned about in America (Hyson 1999, 449). Likewise, the outdoor enclosure of the Africa House did not offer much shelter for the animals, and the exhibit did not differ much from conventional modern concrete structures with radial, arena-like plains. In fact, most visitors were impatient and most of the exhibit designs indeed guaranteed visibility.

Figure 9.4 shows that plants camouflaged the moat from the perspective of the visitors. The trees in the background of the outdoor enclosure, locust trees, were acacia look-alikes, which provided the mood of the African savannah, at least as symbolic references. Bamboo resembled papyrus. The vegetation was not intended to create a naturalistic illusion, but in favor of the visitors, the design of the enclosure departed from well-structured modernism. Plants screened physical barriers, and wooden planks covered the concrete walls inside the enclosure. The wood prevented the rhinos from rubbing off their horns on concrete walls, but the planks also decorated areas that were out of reach for the rhinos. Therefore, these wooden planks fulfilled aesthetic intentions because they distracted from the view of the concrete walls and visually let the enclosure appear more natural.

On the level of representation, contemporary preferences of architectural style and collective images of visual African "nature" influenced the appearance of the buildings and enclosures. The modernist architecture in Zurich seems to convey more rational and honest information about behavioral needs, and about technical and veterinary requirements than earlier nature-illusionistic sceneries had – for example, Hagenbeck's famous Panoramas and their successors in other zoos (Hanson 2002, 152). Hediger had to demonstrate that the maintenance and the exhibition of wild animals fostered scientific and cultural values. Apart from these

Figure 9.4 Exterior enclosure of the Africa House, Zoological Garden of Zurich

Source: Zoo Zürich, 1965.

contemporary representational building codes, the visitor space was designed to provide improved circulation, unobtrusive views and barriers that prevented the visitors from violating the animals' space.

The Return of Naturalistic Environments: The Otter Enclosure

The Ape House represented progressive technical and scientific developments of modernity, and the design of the Africa House expresses aesthetic and cultural values in combination with Hediger's theory of territory, which was transposed into a semiotic program. The otter enclosure, opened in 1970, illustrates the problem of cohabitation, namely the conflict between the Swiss population and native otters, and the territorial conflict between otters, solitary animals, in captivity. Otters were sometimes killed by zoo visitors due to their alleged threat to regional fishery. The design of the otter enclosure was intended to improve the public's relationship to the endangered animals after zoo visitors had already killed three otters at Zurich Zoo (Schmidt 1972, 83). Thus, the otters on display were intended to propagate the importance of wild living otters in Switzerland. Pursuing the conservational approach of zoological institutions that had been promoted since the 1950s, the zoologists planned to reintroduce otters into "wild Swiss habitats" by releasing the expected offspring (ibid.). Landscape architect Albert Zulauf and construction manager Marcel Perrin cooperated with the zoologists to design the exhibit according to the veterinary and behavioral needs of the otters, and with the aim to address the public with a positive image of the alleged vermin by providing special viewpoints. In addition, the otter enclosure had to integrate hygienic features behind the scenes, advancing breeding conditions.

The otter enclosure was set up in the northwest of the Africa House and provided "almost natural conditions" for the otters. A polygonal building structure was integrated into the hillside: each element was based on a hexagonal module (Figure 9.5). The visitors had to leave the main route and enter the dim visitor space, a flat roof viewing structure. The outdoor exhibits were secured with plate glass, which isolated the public from the animals (Schmidt 1972). The building offered different spatial relations between visitors and otters by close-up views of sleeping otters, and by viewing lively, swimming animals in their environmental context from a lower level, which should create a lower position for the visitor. Plants and wooden logs allow the exhibit to appear natural in a mimetic sense. The four enclosures were divided by glass screens and separated individual or pairs of otters (basically solitary animals that might react aggressively when confined together in small spaces). Nevertheless, the planners wanted the public to perceive otters as friendly and playful so that the glass screens helped to convey the impression of one social community in one continuous space. An aquarium with regional fish species was set up on the opposite side of the boxes, peacefully contextualizing the otters and their natural environment.

Figure 9.5 Otter Enclosure, Zoological Garden of Zurich

Source: Pressedienst Diamant Foto, Zürich, 1970.

This manipulation shows how Hediger departed from his earlier intentions to communicate information about animal behavior, especially about animal territories. Instead, the visual illusion of coherent space and social community was an anthropomorphism to improve the public image of otters. Presenting either solitary or fighting animals would have affected Hediger's promotion adversely. The structure of the shelter and the landscaping provoked specific behaviors by the otters as well by the visitors. Viewpoints did not suggest control over the animals, but the viewing structures rather placed the visitors on a subordinate level. To summarize the intent of the display, the conservational message gained much more importance than accurate information about otter behavior. In contrast to the sculptural architecture and the use of solitary plants in the Africa House, the otter exhibit presented a rather illusionary idea of Swiss nature by focusing the views into diorama-like displays in favor of communicating conservation.

On the one hand, the hexagonal modules of the layout corresponded to Hediger's theory of the polygonal territorial space but, on the other hand, they also referred to contemporary popular forms in landscape design. For example, similar polygons were used at the Garden Exposition "Grün 59" on the shore of Lake Zurich and had been used for the zoo's open-air terrariums in the zoos of Zurich and Basel since the early 1960s. But the polygons were also the dominant formal elements of Charles Luckman's master plan for the Los Angeles Zoo and Botanical Gardens, planned in 1959 and opened in 1966 (Hancocks 1971, 137). It is unlikely that these designs resulted from direct correspondence: using polygonal modules had become common for contemporary architecture since the premises of modern architecture changed and led to the breakup of the *Congrès Internationaux d'Architecture Moderne* in 1959. Instead of modern concrete cubes, architects were expected to respect human behavior, and, by doing this, many designed or relied on semiotic interpretations of space. In 1970, Hediger wrote that the zoo's nature was a combination between an animal's and a human interpretation of natural space and prospects:

> Today and in the future, 'zoo' must mean 'zoological garden' since it is a loose and tasteful combination of animal territories, as natural as can be, [...] in order that these integrated spatial portions appropriately transposed from nature at the same time serve as human recreational space.
>
> (Hediger 1971, 8)

Hediger Transformed

Hediger's claim for a better quality of animal cage was deliberately misinterpreted when *Wild Animals in Captivity* was translated into English in 1950. Hyson points out that a review of the book in *Times Magazine* concluded that animals liked what we liked: "Animals were better off in the zoo [...] regular meals, leisure, protection and the friendship of man" (1999, 446). If the zoo's staff entertained the animals, the meager design of their enclosures would not harm their physical and mental health. This interpretation contrasted Hediger's initial intention of avoiding anthropomorphism in zoos because these animal houses resembled modernist apartments. An experimental adaptation of Hediger's concept of relational and flexible space resulted in the draft of the high-rise zoo, a utopian design for a new urban zoo in Boston (Geiger 1968). The vertical zoo consisted of modular spaces and enabled the designers and zoo managers to modify the static materiality of architecture (Figure 9.6). The project reacted to the density of the city center of Boston, and intended to import nature into the city. The structure was essentially cubic, and thus it did not transport the semiotic characteristics of nature, at least not in terms of Hediger's aesthetics of natural space. Nevertheless, it was flexible and adaptable to the changing needs of the users. The Swiss zoo director harshly rejected the project of the progressive architects Cambridge Seven Associates, and insisted on the conventional typology of zoos as landscape parks (Hediger 1971, 6).

Hediger's spatial theories of territorial space, captivity, and flight distance not only influenced animal husbandry, but also concepts for human dwellings. Learning to design sufficient environments for animals implied the reflection on human dwellings. This direct connection is never mentioned in the book *Man and Animal in the Zoo*, although the author stresses the important function of a zoo as "emergency exit to nature" (Hediger 1965, 80), since economic and urban growth severely influenced Swiss landscapes (Pfister 2010). Comparing urbanites to zoo

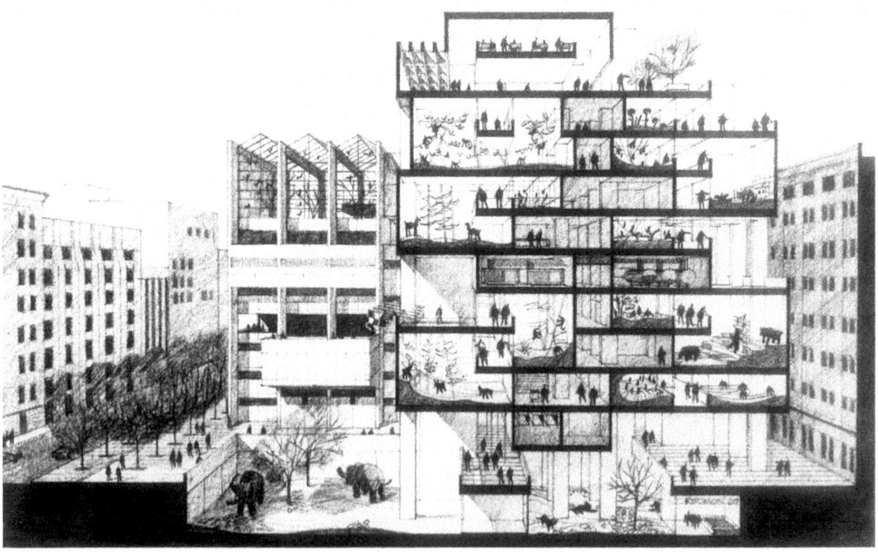

Figure 9.6 High-Rise Zoo, Cambridge Seven Associates. Project for the Boston Zoological Society

Source: Cambridge Seven Associates, Inc, 1968.

animals, several biologists working in zoological gardens investigated pathologies caused by environmental deficits in the 1960s. For example, Desmond Morris, curator for mammals at London Zoo, took up this obvious analogy and published his book *The Human Zoo* in 1969, comparing the pathologies of city-dwellers to those of zoo animals.

The semiotic interpretation of the environment directly influenced planning concepts: Hediger directly influenced the architectural theory of his personal and professional friend Fred Fischer, a medical scientist and director of the Swiss Neutra-Institut, who used the studies on territory and flight distance for analyzing housing and landscaping (Fischer 1965). Typical of the semiotic discourse of architectural theory, Fischer structured basic spatial forms into certain readable codes and thereby transferred Hediger's behavioral studies on animals to the social activities of humans. Still, Fischer's psychological theories for spatial design are valid to explain certain effects of topological features on human inhabitants (Sauter, Hartmann and Katz 2011). The necessity for inventing more complex and secure spaces influenced developments in environmental psychology: Edward T. Hall based his development of proxemics on Hediger's work about flight distance. Both scientists described social territories defined by invisible, psychological barriers (Hall 1959).

While these approaches refer to theoretical considerations about space and communication, Hediger's findings about topological and semiotic aspects of space directly influenced planning practices for the design of mental hospitals (Ellenberger 1974). Working with the architect Kiyoshi Izumi, the psychiatrist Humphry Osmond applied Hediger's concept of relational space – and especially his concepts of flight distance and designation of territory on spatial design (Sommer 2004). Hediger's ideas about qualitative environments helped to comprehend the impact of architecture on the social well-being of its inhabitants. The welfare of patients became the main task of the design, and it paralleled the management of behavior and social life for captive wild animals. Nevertheless, Ellenberger pointed out the need for a reinvention of both institutions (Ellenberger 1974, 59–92).

With regard to the question of the experimental potential of animal-oriented zoo architecture, none of the animal buildings of the Zurich Zoo were amenable for experiments with architectural elements because, on the one hand, the structures were too static and expensive. On the other hand, the buildings, especially the Africa House, were designed as a cultural stage for the animals and a sculptural manifesto of Hediger's thoughts on topological, animal-related forms. As a consequence of the zoo's scientific programs, aesthetic considerations were adjusted accordingly.

Zoomorphism was a means of mediating animal spaces to the visitors and demonstrating the differences between specific needs of species and individual animals. The new zoo architecture in Zurich worked as a representation of knowledge, scientific accuracy and the cultural relevance of exhibiting non-domesticated animals. Therefore, Hediger's writings helped to establish zoo architecture as a specific construction project with specific strategies. Changes in design, scenographic representation and general holding conditions only developed along with changing ideas for human environments and living spaces. Hediger adapted his written statements to the changing styles of the buildings and their interpretations of nature. Accurate studies on built and social environment – and a better understanding of biological relationships and individuals instead of species – enhanced the power of the architects to transform and construct artificial environments. In contrast, the actual practices of husbandry were carefully hidden behind the scenes. The keepers supported the stages of environment and its actors invisibly, like stage technicians.

Notes

1 This article generally refers to the first edition in German. The English translation *Wild Animals in Captivity* was published in London in 1950 and in New York in 1964
2 For example, see Hancocks (1971, 149); Hediger (1964, 43).
3 Apart from brief comments on Hediger in several publications in the field of cultural history of zoos, Röthlin and Müller mentions the buildings in their detailed history of the Zoological Garden of Zurich, and its current director Alex Rübel published Hediger's biography and brief reviews of selected books (Röthlin and Müller 2000; Rübel 2009).
4 For detailed studies on Hagenbeck's scenography, see Rothfels (2002) and Ames (2008).
5 For Haefeli's architectural work, see for example Hildebrand, Maurer, and Oechslin (2007).
6 Photographs and notes still exist in the Archive of the Zurich Zoo, Files 1954.
7 Planning material of the general plan is located in the Archive of the Zurich Zoo, Files 1956 "Zooausbau."
8 Dr. René Honegger, the former assistant to Hediger and curator at the Zurich Zoo, and Manfred Studer, the former head rhino keeper, contributed helpful insights to this chapter.

References

Anonymous (1959) "Architektur für Affen," *Werk-Chronik*, 11: 227–228.

Ames, E. (2008) *Hagenbeck's Empire of Entertainment*. Seattle: University of Washington Press.

Baratay, E. and Hardouin-Fugier, E. (2004) *Zoo: A History of Zoological Gardens in the West*. London: Reaktion Books.

Burt, J. (2002) "Violent Health and the Moving Image: The London Zoo and Monkey Hill," in M.J. Henninger-Voss (ed.), *Animals in Human Histories: The Mirror of Nature and Culture*. Rochester, NY: University of Rochester Press, pp. 258–292.

Burt, W.H. (1943) "Territoriality and Home Range Concepts as Applied to Mammals," *Journal of Mammalogy*, 3: 346–352.

Chrulew, M. (2013) "Preventing and Giving Death at the Zoo: Heini Hediger's 'Death Due to Behaviour'," in J. Johnston and F. Probyn-Rapsey (eds), *Animal Death*. Sydney: Sydney University Press, pp. 221–238.

Ellenberger, H.F. (1974) "The Mental Hospital and the Zoological Garden," in J. Klaits and B. Klaits (eds), *Animals and Man in Historical Perspective*. New York: Harper Collins, pp. 59–92.

Fischer, F. (1965) *Der Wohnraum*. Zürich: Artemis.
Geiger, M. (1968) "Der Zoo im Hochhaus," *Werk*, 4: 219–221.
Gruffudd, P. (2000) "Biological Cultivation. Lubetkin's Modernism at London Zoo in the 1930s," in C. Philo and C. Wilbert (eds), *Animal Spaces, Beastly Places: New Geographies of Human-Animal Relations*. London: Routledge, pp. 222–242.
Hall, E.T. (1959) *The Silent Language*. Garden City, NY: Doubleday.
Hancocks, D. (1971) *Animals and Architecture*. New York: Praeger.
Hanson, E. (2002) *Animal Attractions: Nature on Display in American Zoos*. Princeton, NJ: Princeton University Press.
Hediger, H. (1942) *Wildtiere in Gefangenschaft*. Basel: Schwabe.
Hediger, H. (1949) "Für Tiere Bauen," *Idealprojekt über den Ausbau des Zoologischen Garten Basel. Jubiläumsausstellung*. Basel [s.n.], pp. 22–25.
Hediger, H. (1950) *Wild Animals in Captivity*. London: Butterworths.
Hediger, H. (1964) *Wild Animals in Captivity*. New York: Dover.
Hediger, H. (1965) *Mensch und Tier im Zoo: Tiergarten-Biologie*. Zürich: Müller.
Hediger, H. (1970) *Man and Animal in the Zoo: Zoo Biology*. London: Routledge.
Hediger, H. (1971) "Wandlungen der Zoologischen Gärten," *Anthos*, 10: 1–9.
Heinsdorff, H. (1970) "Bauten und Anlagen zoologischer Gärten," *Deutsche Bauzeitung*, 5: 921–962.
Hildebrand, S., Maurer, B., and Oechslin, W. (2007) *Haefeli Moser Steiger*. Zürich: gta.
Hyson, J. (1999) "Urban Jungles." PhD dissertation, Cornell University, Ithaca, NY.
Kohlstedt, S. (1996) "Reflections on Zoo History," in R. Hoage and W. Deiss (eds), *New Worlds, New Animals: From Menagerie to Zoological Park in 19th Century*. Baltimore, MD: Johns Hopkins University Press, pp. 3–7.
Kupper, P. (2012) *Wildnis schaffen*. Bern: Haupt.
Morris, D. (1969) *The Human Zoo*. New York: McGraw-Hill.
Mullan, B. and Marvin, G. (1999) *Zoo Culture* (2nd edn). Urbana: University of Illinois Press.
Neutra, R. (1956) *Life and Human Habitat*. Stuttgart: A. Koch.
Pfister, C. (2010) "The 1950s syndrome," in F. Uekötter (ed.), *Turning Points in Environmental History*. Pittsburgh, PA: University of Pittsburgh Press, pp. 90–118.
Röthlin, O. and Müller, K. (2000) *Zoo Zürich: Chronik eines Tiergartens*. Zürich: NZZ.
Rothfels, N. (2002) *Savages and Beasts: The Birth of the Modern Zoo*. Baltimore, MD: Johns Hopkins University Press.
Rübel, A. (2009) *Heini Hediger 1908–1992. Tierpsychologie – Tiergartenbiologie – Zoodirektor*. Zürich: Beer.
Sauter, H.M., Hartmann, A., and Katz, T. (2011) *Einführung in das Entwerfen*. Wiesbaden: Vieweg and Teubner.
Schmidt, C.R. (1967) "The Africa House at Zurich Zoo," *International Zoo Yearbook*, 1: 62–66.
Schmidt, C.R. (1972) "New Otter Exhibit at Zurich Zoo," *International Zoo Yearbook*, 1: 83–85.
Steiner, H. (2003) "For the Birds," *Grey Room*, 4: 5–31.
Sommer, R. (2004) "In Memoriam. Humphry Osmond," *Journal of Environmental Psychology*, 2: 257–258.
Turovski, A. (2000) "The Semiotics of Animal Freedom: A Zoologist's Attempt to Perceive the Semiotic Aim of H. Hediger," *Sign System Studies*, 28: 380–387.
Uddin, L. (2015) *Zoo Renewal: White Flight and the Animal Ghetto*. Minneapolis: University of Minnesota Press.

Zoo Zürich (2015) "Mit spitzem Maul und einer Nase (fast) ohne Horn." Press Release. www.zoo.ch/documents/PANO_201501_JungtierNashorn.pdf. [Accessed 25 January, 2015].

10 Care and Tinkering in the Animal House

Conditioning Monkeys for Poliomyelitis Research and Public Health Work

Tone Druglitrø

Introduction

During the 1960s, monkeys started to arrive at the Virus Laboratory at the National Institute of Public Health (NIPH) in Norway in large numbers. Most of them were macaque monkeys: Rhesus macaque (macaca mulatta) and the long-tailed macaque (macaca irus). The macaques had been shipped all the way from India via Copenhagen to Oslo and were to be employed in the ongoing work to identify and diagnose poliomyelitis virus as well as the testing of poliomyelitis vaccines as part of the Norwegian government's public health strategies to combat the spread of this paralyzing and deadly disease. The Virus Laboratory had been established a year earlier, and was presented by the government as a "weapon" to fight viruses by means of diagnosing virus diseases, and by collecting and growing materials that could be used for the production and safety testing of vaccines, in particular polio vaccines that reached the market in the late 1950s (Druglitrø 2012).

In this chapter, I trace the integration of macaques as experimental organisms for polio in Norway during the 1960s and 1970s. Investigating the integration of macaque monkeys helps us to craft an understanding of how political and scientific strategies have been constructed, not only in the shape of various governmental institutions built to tackle epidemic diseases and facilitate public health services, but also, simultaneously, in the bodies and lives of "third world" animals serving as surrogates for "man," to paraphrase Donna Haraway (1989, 2). The aim here is not so much to document the different ways in which the macaques were employed in polio work, but rather to pay attention to *the practices and infrastructures of care* for conditioning the macaques, their organs and tissues as experimental organisms for polio research. These practices were thus crucial in order to "prepare" the macaques for the polio work.

While most of the monkeys that came to the Public Health Institute would be killed for their organs, others were meant to assist in the work of tracing the effects of the vaccines, and some, it was hoped, would reproduce in the animal house and provide new and better-controlled primate bodies. Bringing macaques into the animal house depended on the abilities of the personnel to "condition" the monkeys to the new environments. The practices of conditioning involved making sure that the macaques stayed disease-free and alive as long as necessary, and ensuring that no one was infected or injured in the monkey–human interface: it depended on the development of specific practices of, and infrastructures for, enabling care. Thus, the integration of macaques depended on care practices that could ensure the *co-conditioning*, so to speak, of organisms, people, and infrastructures in order to prepare the necessary materials for combating polio in the laboratory.

My analysis is based on archive materials such as correspondence between scientists, policymakers, animal dealers, and others making a claim on the primates, as well as on scientific publications and media coverage. By treating the various texts as events in themselves, as situated interferences and as active in constructing the contexts (Asdal 2012) for the work to combat polio, it is possible to trace the interface between macaques and humans and the transformations that occurred in the animal house during almost twenty years of monkey husbandry. My analysis will be developed in three interrelated parts: The *first* part discusses recent literature on animal models and experimental organisms, and how the various uses of animals in science must be seen as intertwined with the techniques, methods, and theories available. Following Clarke and Fujimura (1992), the "rightness of a tool" in science is highly situated, local, and historically specific and needs to be investigated by empirical study. The *second* part investigates the significance of reliable procurement practices in developing infrastructures for the use of experimental organisms in science. The import of macaques of unknown origin was in many ways an extraordinary move by the Norwegian science community – a community with neither the skills nor the infrastructures to house such animals, nor a virological research tradition. However, scientists and policymakers across the globe presented the polio epidemic as an extraordinary situation that called for extraordinary measures. The case of the macaques is in many ways unique, but at the same time it provides an interesting case in understanding the different ways in which living organisms have entered the various sites of science. In the *third* part, I move into the animal house and trace the different care strategies that were employed to condition the macaques in the animal house. The analysis demonstrates how scientific work is both made possible and constrained by the access to tools and infrastructures.

The Right Organism for the Job?

In the expanding literature on animal models in contemporary and historical studies of science it has been argued that the work of constructing so-called model organisms has depended critically on building infrastructures around each organism (Ankeny and Leonelli 2011; see also Asdal 2014). The emergence of model organisms has been linked to certain technological and infrastructural developments in the life sciences during the twentieth century (Clarke 1987; Rader 2004) and particularly during its latter half (Clarke and Fujimura 1992; Thompson 2013). In Norway, the work to standardize specific animal species such as mice and rats to become model organisms in public health science was initiated on a national scale in the early 1950s (Druglitrø 2012). When the macaques came to the animal house in the 1960s, the animals living there were mostly small rodents such as mice, rats, guinea pigs, and rabbits. Some of these animals were bred according to principles of genetics and in combination with a controlled pathogenic composition; others were bred according to conventional techniques of genetics and referred to as "conventional animals" (ibid.).

In this context, the macaques procured for polio research form a special case in the history of experimental organisms and animal models from the 1950s. They were not bred for use in science, but caught and trapped outside of the laboratory and the animal house. The unknown origins and biological history of the macaques meant that to build infrastructures and develop skills to condition them for science work was costly, time-consuming, and at times dangerous. This makes it particularly interesting to pay attention *to the efforts to house, condition, and care for* wild monkeys as they created tensions in the animal house because they were *not bred or purpose-bred, and known by various practices of breeding and domestication.*

The purpose-bred animals that were increasingly finding their way into the animal house fit in some ways the definition of a model organism as suggested by Ankeny and Leonelli (2011), in that they were results of agreed and shared standards, and they had defining, generic features. According to Ankeny and Leonelli, model organisms are constructed to be versatile in terms of cross-disciplinary application. In contrast, they claim, experimental organisms are models for specific phenomena to be investigated through a particular discipline with a specific set of techniques and practices (Ankeny and Leonelli 2011, 319).

The macaques that came to the animal house at the Virus Laboratory at the NIPH, however, do not easily fit the definition of either model organisms or experimental organisms. They were not scientifically and technologically designed for a specific purpose; they were procured to be used mainly for polio, but they were also tinkered with to learn various techniques in virology. They were biological containers of valuable materials, and they served as live, biological models for physiological processes. Scientists at the Odontology Faculty in Oslo were eager to experiment on primates to be able to test new equipment and techniques for managing dental hygiene, and parts of the shipment of macaques were distributed from the Public Health Institute to the Odontology Faculty. Rather than pointing to engineered characteristics of the animals, the scientists emphasized the importance of the biological kinship of monkeys to humans. The investigations of the polio viruses (three main strains were identified in the first half of the 1900s) that escalated from the beginning of the 1900s, had demonstrated that these viruses could only cause infections in humans and nonhuman primates (Oshinsky 2010, 208; Guerrini 2003, 117). In this context, the work to identify and trace how the virus produced disease in bodies had to be done on either humans or monkeys. Embryos from human fetuses were used by Albert Sabin to produce and test his live-virus vaccine, and the production of polio vaccine depended on massive clinical trials on humans that sometimes led to devastating results.[1] As such, the macaque and other primate species came to be regarded as next-to-perfect model organisms to understand and to combat the spread of the polio virus. Still, the biological closeness of nonhuman primates to humans was double-edged: nonhuman primates offered great potential in terms of producing human health benefits but at the same time they posed great threats in terms of transporting life-threatening diseases to humans in the monkey–human, monkey–science interface.[2]

However, the rightness of the macaque for polio work was not only a matter of biological kinship and compatibility, or one of ethics; it was also a matter of the techniques, methods, and theories available to the scientists at the time. Intact primate bodies had been extensively used in polio research since the beginning of the twentieth century, as scientists believed that the virus could only propagate in tissue or organs inside living primate bodies (*in vivo*). This changed with findings in the late 1940s, that the rhesus monkey (macaca mulatta) kidney cell could be used for cultivation of human viruses *in vitro* (Enders et al. 1954). By using penicillin to avoid infection of the tissues by bacteria, tissue culture techniques were regarded as a more efficient procedure in order to produce enough polio viruses to experiment on and to make vaccine production practical. From the 1900s, monkeys – particularly rhesus monkeys – had been used to identify polio viruses as well as to trace how the polio viruses affected primate bodies. With the development of tissue culture techniques, primates were used to collect cell and tissue cultures that could be used to cultivate polio virus for vaccine production. The main function of the macaques when they came to the Virus Laboratory was to provide organs and other organic materials. To understand the central difference between the work to condition the macaques that initially came from unknown sources in Asia and the animals that were bred by certified breeders in certified sites, we need to start the analysis before the macaques' arrival, and trace their route to Oslo and the NIPH.

Procuring Macaques

A central challenge to using macaques was the unreliable and uncertain character of the supply chain. While purpose-bred animals were either procured from local breeders, over which the scientists had some control and could inspect, or from large-scale producers abroad that were controlled and regulated, also by the science community, the macaques originated in the jungle or in the urban vicinities in India and Indonesia. When the animal house was built at the Virus Laboratory in Oslo, the first task was to set up reliable domestic contract breeders or to import animals of different species from certified institutions abroad. In 1959, Dr. Oskar Lahelle, the head of the then newly completed laboratory, applied for approval from the Ministry of Agriculture to import primates from India. The Ministry complied, but demanded that Lahelle obtained documentation by Indian authorities on the health status of the monkeys. Even though macaques were increasingly employed as laboratory animals, procuring them was not easy (Krag 1960). A central concern for the scientists was that the primate trade was in many cases a shady business, and not easy to control. This was also pointed out at international symposia during the 1960s and 1970s on the use and breeding of primates, in discussions on how to get control over the primate supply chains. The sudden increase in demand for primates for science had placed great strains not only on the natural habitats of primates but also on the quality of the care taken to ensure that primates survived the transit and stayed healthy (Schmidt 1972, 2).[3]

Thirty monkeys were expected to arrive in the first shipment to the NIPH, followed by an estimated twenty-five monkeys every second month during the first year. The number was expected to increase in the following years (Lahelle 1959b). Lahelle had investigated different sources of primates; he contacted animal dealers in Denmark and in Holland. The choice of supplier for the NIPH was Carl H. Krag, the same animal dealer who was serving the State Serum Institute in Copenhagen.[4] Lahelle wrote to the Norwegian government about the arrangement:

> This is to certify that [NIPH] has authorized the animal dealer Mr. C.H. Krag, in Denmark. [...] The monkeys are intended for testing of polio vaccine and preparation of tissue cultures. They will be carried by air and arrangement will be made to ensure their safe transit under human conditions.
>
> (Lahelle 1959a)

In December 1959, a shipment of monkeys arrived at the animal house. In line with Norwegian custom restrictions on the import of animals at the time, the primates were to be isolated immediately at the institute, and to stay there until their bodies were cremated after use (Baggerud 1959; Lahelle 1959c). The macaques that came to the animal house at the NIPH had started their trip in India, where they had been caged and prepared for shipping by plane to Kastrup airport in Copenhagen, Denmark. On arrival in Denmark, the monkeys were kept in their containers or they were housed by Krag, until they were shipped to Oslo and the NIPH. In transit they might receive water and some food, but they were kept in their cages, because if released they could pose risks in terms of spreading infection and disease. The monkeys were thought to be potential carriers of a range of different parasites and viruses originating in the jungle, or most often contracted in the unhygienic conditions during transport.

The long journey was not only uncomfortable and possibly frightening for the monkeys; the trip also posed a serious threat to their health. The encounters with different people, pathogens, and cage technologies on their route from India to the NIPH, were crucial moments in determining the primates' potential for survival. It was, crudely speaking, up to the macaques themselves to survive the trip to the animal house.

In his detailed descriptions of the practices of trapping, caging, and transporting nonhuman primates from Asia and Africa to Europe and the US during the 1950s and 1960s, psychiatrist W.T. Roth reveals the many perils the monkeys were exposed to:

> The challenges to which a newly captured primate must adapt to survive are formidable. The captive is removed from his social and natural environment and plunged into an upsettingly unfamiliar situation. He receives unfamiliar feed and water, is held together with unfamiliar cage mates, is confined in a succession of unfamiliar surroundings and subjected to a variety of rapid climate change.
>
> (Roth 1969, 7–8)

Even though care was taken in terms of providing separate cages for female and male macaques, or providing a bowl of water, the monkeys would often respond to captivity negatively; that is, they could show signs of illness, or die in transit.

Tinkering with the Macaques: Developing Practices of Conditioning

Compared with the manner in which the macaques were brought into the animal house, where they had been caged and, to various degrees, been taken care of, the housing of the macaques at the NIPH involved a range of technological installations and architectural adjustments. Hence, to be transformed into organisms "ready" to combat polio, the macaques had to be *conditioned*; they had to be habituated and prepared in the new infrastructural arrangements in a manner that would ensure health and life.

In STS, feminist perspectives on the concept of "care" is proposed as a lens through which power relations can be analyzed differently than as top-down affairs. The concept of care suggests the investigation of interactions in a more open-ended way, as processes of constant negotiation between humans and nonhumans, where the relations of power are not given, but established in practical *situations*: To care for others is experimental and contingent to situations (see Mol, Moser, and Pols 2010).[5] The use of *tinkering and care* in this chapter is meant to capture the work of conditioning the macaques, and to demonstrate how it depended on the constant tinkering with animal and human bodies and the technologies and material arrangements in which they interacted.

Such an approach makes it possible to tease out how the infrastructures for polio work depended on the distributive relationship between animal technicians and animals, as well as the distributive relationship between animals and the animal house. The mutual conditioning of macaques, technologies, and humans depended on tinkering that involved *care* in specific forms. However, is *care* the appropriate way to conceptualize the practice of conditioning wild monkeys in the animal house? Does it work as a euphemism, or does it aid in understanding how relations of care in their various forms can be fused with tensions and contradictions, and as such help us avoid falling into the trap of pre-judging "good" versus "bad" practices? I propose that the concept of care does not negate the hardship of monkeys and humans in the animal house, but helps us recognize how monkey husbandry played a crucial role in keeping aligned the activities of public disease control, the scientific laboratory, and the animal house (for example Lewis et al. 2012, 780). Further, while the concept of "tinkering" is defined in dictionaries as efforts to repair or improve something in a casual, indiscriminate, and unfocused way, I propose that in the context of the animal house, in which the development of knowledge and skills on animal husbandry was the main task, tinkering was far from unfocused or indiscriminate; tinkering was an integral part of the "becoming with" of animals, humans, technologies and the very animal house itself. As such, tinkering captures the work in the animal house as being an activity that is partially calculated and partially about searching for good solutions, in this case in an extraordinary situation of a viral epidemic. Hence, when coupled with the concept of care, tinkering is useful to capture *how the form of care* that was needed shifted constantly between attention to the animals, themselves as caretakers and agents of science, and the infrastructural arrangements.

The first job of the staff at the animal house was to make sure that as many as possible survived the transition from the outside to the inside of the animal house. Often when entering the house, macaques would die or fall sick within two or three days after arrival. Sometimes two or three months could elapse before the macaques started to show signs of disease. In the in-house bulletin, the head of the animal house wrote:

> In 1964 we received 4 different shipments with a total of 43 animals, every one macaca irus (cynomolgus). Unfortunately we experienced substantial loss among these animals. In total 13 were lost, most of them during the first weeks after reception. This is obviously a tough time for them with an abrupt transition from a life in freedom to a life in captivity under which the environmental, nutritious and climatic conditions are radically changed.
>
> (Erichsen 1966, 13 [my translation])

The uncertainty about the various viruses and bacteria to which the macaques had been exposed and might carry from the point of origin or in transit, made it crucial for the Norwegian scientists to conduct investigations to learn how to respond preventatively to the so-called "built-in" liabilities emphasized in various international publications on the housing and use of primates (Schmidt 1972, 1). Here, tinkering and attentive experimentation provided a foundation for performing care of wild primates in the animal house:

> Investigations have shown that the number of animals with B-virus antibodies in the blood is low when the animals enter the laboratory, but during housing they can increase considerably. This shows that the infection can be present in the stock by single animals when captured, and that it will spread fast to non-infected individuals under the conditions in which they are in entrapment. All of this occurs usually without any signs of illness in the animals.
>
> (Erichsen 1963 [my translation])

The task of the staff at the animal house then was to try to keep the built-in liabilities of the macaques in check. The built-in liabilities could manifest themselves either as disease in the macaques, disease in other animals housed at the animal house, or in the people working there. Further, the built-in liabilities could be provoked by the macaque–human, macaque–animal-house interface. The animal house and/or the trip to the animal house was in fact identified as an imperative factor in causing illness and death of the macaques: "We get the impression that parts of the losses are linked to maladjustment, as the autopsies only show extreme emaciation without organ alterations" (Erichsen 1965). In the following years, Erichsen reported identical incidents (1966, 1967). It was suggested that the strict, hygienic regime that awaited the primates, and that were constructed to keep dangerous pathogens in place and to care for the monkeys, might be working against its purpose by provoking disease and death. The elusiveness of the macaques – the range of unknowns that accompanied them – thus demanded tinkering; it demanded attentiveness and even anthropomorphism.[6]

Conducting such investigations was regarded as risky, as most infections contracted by workers in animal houses had been reported to occur in contact with monkey substances and not live monkeys, such as the handling of blood and tissue cultures, or during autopsies and other surgical interventions, or in the experimental work of preparing tissue cultures. An incident at the animal house in the early 1970s illustrates how the macaques' bodies were regarded as risky, carrying potentially dangerous infections that were provoked or revealed by surgical interventions:

> The time of delivery is always long, but they usually arrive as expected, so that the users are covered, even with six weeks of *conditioning*. However the biggest problems in deliveries we had was due to an unusually high frequency of "foamy agent" infection[7] among the macaque monkeys. In total we lost 14 monkeys and as usual most of them just after arrival. For the first time in years we had in 1971 seven cases of tuberculosis, in a shipment of macaca mulatta (rhesus) brought in for a special purpose. The unusual aspect of the outbreak was that the type of bacteria showed itself to be Typus bovinus (bovine tuberculosis). This could be the reason that the animals supposedly were recorded to be tuberculin-negative at the supplier. However, tuberculin testing at the department revealed more infected animals that were all immediately killed. In other autopsies one case of Shigella infection[8] was identified. More cases of Shigella infection were also found during clinical investigations.
>
> (Erichsen 1972 [emphasis added, my translation])

Such investigations demanded tinkering and were central to the processes of conditioning of the macaques, as it could aid the staff to relate the outbreak of diseases or the death of a macaque to potential sources such as the animal house itself, the animal dealer, or to links earlier in the supply chain. The animal dealer would sometimes pay compensation for dead macaques, however only if deaths occurred just after their arrival (Erichsen 1967). If time elapsed before macaques started to die, the causes could not as easily be traced back to the animal dealers, and could only be explained by arrangements and practices of care in the animal house. As such, the investigations offered indications of unwanted pathogens circulating in the animal house.

Hence, observing the macaques' ability to adjust to the new environments and investigating their bodies by performing necropsies and histological inquiries, was part of the attentive experimentation – the tinkering – necessary to learn and assess potential causes and risks. Moreover, it was central to the work of "conditioning" the macaques into the new infrastructural arrangements as it depended on the development of skills to improve housing conditions. These improvements were increasingly materialized to enable care that could be part of the process of conditioning of the macaques.

Enabling Care and Handling Risk in the Animal House

An entire floor was set aside for the macaques so that they were separated from the other animals in the animal house. The "monkey floor" consisted of three monkey rooms where the macaques were separated by gender, species of monkey, and purpose. In each room the macaques were separated in steel pipe cages.[9] The practices of separation were regarded as central to keep the monkeys apart from other animals, as high concern was expressed regarding the spread of virus from monkey to monkey, monkeys to other animals, and from monkeys to humans through clothing, substances, waste, and so forth. Placing the macaques in single cages was part of the effort to control them and their interactions with the other macaques, and to prevent incidents of collision between them. Care in this context was not about nurturing the species specificity of the macaques as such, but was rather a way of conditioning the macaque bodies by stimulating their "potential" as organisms and materials that scientists could use for prescribed purposes (Friese 2013, 130). Thus, rather than being oriented directly towards the macaques, a major part of the work of caring for them was about *enabling care*, that is, to maintain the infrastructures of the animal house, to install suitable technological equipment, and to tinker with various solutions to ensure that the atmosphere in the animal house complied with necessary standards of hygiene.

Let us look at how the infrastructural arrangement of the animal house changed as a result of tinkering during the 1960s – a transformation that can be regarded as processes of co-conditioning of humans, technologies, and animals. In the first half of the 1960s, it was reported that it was "impossible to maintain an acceptable level of hygiene" in the monkey rooms, and the technological equipment was "primitive" (Erichsen 1969). Controlling the flow of viruses and pathogens was reported to be difficult, and so was the cleaning of the rooms. By the late 1960s, a high-pressure washer was installed that was said to work perfectly, and which kept the rooms in better shape. Other technologies were tested that could be used to ensure a sterile airflow in the primate rooms. For instance, after a few years of housing macaques, UV lamps were installed to kill micro-organisms residing in the monkey rooms. Adjustments were made to the structures as well, such as decorating the walls with tiles and changing the flooring from paint that needed regular rehabilitation to a more durable epoxy that made it easier to clean and maintain the monkey rooms. It was reported in the in-house bulletin that the changes placed the rooms in line with international standards (Erichsen 1968a; Erichsen 1970).

The changes in the architectural infrastructures and integration of new technologies were the results of tinkering in the animal house and caring for the macaques. The changes were also a result of tinkering with other experimental animals housed at the animal house, and practices of care were often translated from one species to another. Importantly, the various arrangements and technologies in the animal house were meant to benefit the work of the staff – to make it easier and more efficient. Most of the work at the animal house was linked to the routine work of the various laboratories, such as the tissue culture laboratory and the histological laboratory, and as such the organization of the animal house had to be synchronized with the pace and preferences of these other sites. The infrastructural arrangements organized the complexities of the care work.

For instance, caring for the macaques and the animal house involved handling a range of risks. The arrangements enabled the staff to distance themselves from the macaques but at the same time move closer to them. Even though most of the members of staff were not in immediate contact with the macaques, a range of precautions and instructions were given for the well-being of the staff, the well-being of the monkeys, and the "well-being" of the animal house by preventing risky situations. There were instructions for what to wear, procedures for procurement and sale of animals, animal accounting, killing of animals, how to treat sick or dead animals, the handling of waste, how to use and clean cages, how to use and clean water bottles, temperature and so forth. The function of these quite varied but detailed instructions was to prevent the outbreak of disease in individual animals and animal stocks. Preventing disease had thus both an economic and a qualitative aspect to it, and depended on efficiency and skill to work as successful strategies. The instructions told the personnel how to move around in the animal house by giving assigned tasks to specific positions, as well as directions on what to wear and how to conduct oneself safely. The tasks could range from the cleaning of cages, rooms and instruments, to recording incoming and outgoing animals, as well as how many animals died and how many were born (even though breeding was not a central task at the animal house of the NIPH).

Even though only specified personnel were allowed in the monkey rooms, it was mandatory for all personnel at the animal house to be vaccinated to ensure immunity against different viruses that could be transmitted from the monkeys:

> The leap from a mouse bite to a monkey bite is serious. I am sorry to admit that people have lost their lives after being bitten by a monkey because the monkeys are carriers of the B-virus. These accidents have occurred in larger institutions abroad where the dimensions of the primate stocks have been much greater. The more primates one takes in, the greater is the chance to get

> animals that are carriers of disease. These will over time contaminate their comrades. [...] One has to take as a point of departure that all, or some of them, can be dangerous at one point or another during the infections. In our little unit the risk is not as serious, however we still need to count on it and take precautions to protect ourselves.
>
> (Erichsen 1963 [my translation])

Furthermore, instruction number 17, which was directed specifically at monkey husbandry, read that staff had to wear special protective clothing such as rubber boots, coats and face masks when entering the monkey rooms. Instruction number 5 gave instructions on how to clean cages. Both the cleaning of the monkey cages, the monkey rooms, and the handling of the monkeys themselves involved a lot of hard and risky work, and the protective clothing and face masks were central to avoid the transmission of disease as well as injuries (Erichsen 1968b [my translation]).[10] Instruction number 5 read:

> Manure pans and grates are to be cleaned in the monkey room by flushing them with iodine solution from apparatus. Mouth-nose masks and face protection are to be used during this operation as one has to expect that dangerous microbes can spread into the air during the cleaning. Cups for food and drinking are to be disassembled once a week, soaked, washed, and steamed in the monkey section's own washing room. If cages or parts of the cages are to be moved to other sections in the department they always have to be sterilized in an autoclave at 120 degrees [Celsius] for 30 minutes before being moved.
>
> (Erichsen 1968b [my translation])

Disposable plastic bags would later replace the manure pans so that the manure would not be lying in the pans uncovered, because manure pans that were not covered or secured in closed bins were thought to result in bad hygiene, smell and general atmosphere in the animal rooms. Another instruction targeted monkey husbandry in particular, and read that:

> Monkey bites or skin abrasions, and other wounds incurred in the monkey rooms, must be rinsed immediately with 3% hydrogen peroxide. ... All injuries incurred in the monkey rooms or during work with monkeys, monkey tissue, monkey blood, materials from the monkey rooms and the like are to be reported immediately to the head of the department or his deputy.
>
> (Erichsen 1964 [my translation])

Thus the conditioning practices revolved around a varied set of risks, in which strict instructions were given on how to behave and work; they took part in structuring the work of the humans working at the house (such as caretakers, technicians, and cleaners) as it depended on organized teamwork. At the same time the handling of these risks depended on continuous attentive experimentation of these various relations of risk.

At this point, I would like to return to the question posed earlier: Is "care" the appropriate concept to describe the activities of integrating macaques in the manner done during the polio epidemic? On the one hand, care and tinkering has helped me tell this story, and to place particular focus on the relationship between animals as experimental organisms and animals as wild nature, between infrastructures and care work, and between bodies and technology. On the other hand, the concept of care becomes demanding, as the strategies of housing the macaques were just as much about the innovation and organization of technologies, and the care for the health and welfare of the staff, as it was about caring for the macaques in their own right. In one sense, the conditioning of macaques can be described as practices of caring for the elusive pathogens and viruses that would potentially accompany the macaques into the animal house and pose threats to science, other laboratory animals, and humans. Moreover, the members of staff seem to have learnt more about different strategies for housing and conditioning animals and how to handle infections than about the macaques themselves.

From Unconditional Monkeys to Conditioned Monkey Parts

"Once mastery over particular research material has been achieved, future work is constructed," Clarke and Fujimura write (1992, 9). It would be an overstatement to argue that the scientists and people working in the animal house at the Virus laboratory managed to gain mastery over the macaques coming from India during the 1960s and 1970s. The last reported use of live monkeys in Norway was in 1978. The use of monkeys had decreased during the 1970s, and from the mid-1970s, around fifteen monkeys had been imported annually until the last reported import and use of live macaques at the NIPH in 1978. This was the same year as India and Bangladesh placed an embargo on the export of primates to Western medical science and pet industry.

The NIPH in Oslo started to import refrigerated monkey organs and blood from Sweden to produce antibodies in tissue culture, which again could be used to diagnose polio virus as well as test and produce polio vaccines. It was only parts of the monkeys – that is, only body parts and organs – that were integrated into Norwegian virological research and diagnostic work. The controlled procurement of monkey body parts and substances from a reliable distributer was less time-consuming, expensive, and risky.[11] The issue of economic cost was central, but so was the opportunity to import primate materials that were safety tested by reliable suppliers and worked as reliable experimental organisms.[12]

The science community debated the need for a Scandinavian breeding unit of monkeys to supply virological, odontological, and immunological research and public health work, however they never dreamed of establishing a national unit in Norway (Erichsen 1975). These deliberations were integral to international developments on the standardization of animals for science and the emergence of a market for the production and distribution of laboratory animals in the same period. At the European Symposium on the Use of Non-Human Primates in

Medical Research (under the auspices of WHO) in 1967, the issue of the large-scale breeding of primates and the need to find a standard primate model was presented as urgent concerns for the development of biomedicine and public health. These discussions continued into the 1970s, in which it was claimed by the science community that "positive steps must be taken to develop sources of supply under the control of the end user." Positive steps involved the establishing of breeding colonies of primates and facilities on scales that were "designed to meet major needs", combined with being affordable for the users (Schmidt 1972). They were termed positive in that the scientific community would no longer be dependent on the draining of natural resources of primates from the wild, and that the monkeys used would be purpose-bred and consequently *known* to the scientists.[13] Breeding monkeys would mean that they were able to circumvent the many risks, uncertainties, and issues concerning the material quality of primates for science and of stable and reliable supply chains. The economic aspect was important not only in terms of restricting the costs of animal-based science, but was intertwined with the characteristics of public health strategies dependent on efficiency and economics. The changes in the practices of procuring monkey body parts meant that macaques from the wild would no longer be housed at the animal house at the NIPH; the animals remaining in the animal house were mostly purpose-bred, of and for use in the laboratory.

Notes

1 For more on the clinical trials of polio in the 1950s and the infection and death of several children as a result of the trials, see for example Guerrini (2003).

2 A series of international symposia was arranged on the issue of nonhuman primates in science that also discussed widely the potential dangers of housing them as well as experimenting on them due to the transmission of disease from nonhuman primates to humans. Vaccinologist and pharmacologist Schmidt stated in 1972 at an international symposium on the breeding of nonhuman primates for laboratory use in Berne the year before, that: "On occasion wild-caught monkeys and apes present hazards to the well-being of other primates, including man, and thereby jeopardize unrelated investigations" (1972, 2).

3 People involved in dealing in monkeys in India and other Asian countries were originally tradesmen from other commercial fields lacking experience in the animal trade: "It took them years to learn some of the basic requirements inherent in dealing with live wild caught animals and to develop improvements" (Roth 1969, 3).

4 The animal house at the NIPH was already importing smaller animals from the Serum Institute (for more on this, see Druglitrø 2012).

5 See also Timmermans and Berg (1997) and Haraway (2008).

6 See Law and Lien (forthcoming), for a discussion on the issue of relational anthropomorphism.

7 The first description of foamy virus (FV) was in 1954. It was described as a contaminant in primary monkey kidney cultures. The first isolate of the "foamy viral agent" was in 1955.

8 Shigella was said to cause disease in human and nonhuman primates, but not in other mammals.

9 What types of cages to be used was discussed at length, and the staff contacted colleagues abroad to hear about their experiences on different types of caging. Price

was also central to what kind of cages they chose to buy; they needed to be robust enough to endure continuous use and cleaning.

10 Caging systems were also installed in the animal house, meant to make the work in the animal house more efficient. The cages were now hung on rails from the ceiling and mounted on the walls, making it easier for the technicians or the cleaners to move them around.

11 Anita Guerrini writes that the use of tissue culture techniques meant that one monkey, or a monkey kidney cell strain, could provide two hundred culture tubes, as opposed to a single monkey spine resulting in three samples of virus (2003, 122).

12 The use of monkeys also reached the headlines of the newspapers in the mid-1970s, with the rise of animal rights advocacy and the ongoing processes of developing a new Animal Protection Act (1974), and had an effect on the changes in import practices.

13 Organizations such as the International Council for Laboratory Animals (ICLA) aimed to establish standards of various species of animals for use in biomedical research and emphasized the importance of this work for the further development of medical science (Druglitrø and Kirk 2014).

References

Ankeny, R.A. and Leonelli, S. (2011) "What's so special about model organisms?," *Studies in History and Philosophy of Science*, 42: 313–323.

Asdal, K. (2012) "Contexts in Action: And the Future of the Past in STS," *Science, Technology and Human Values*, 37 (4): 379–403.

Asdal, K. (2014) "Contesting the Animal Model: Axel Holst and the Controversy over Scurvy and Beriberi," *Social History of Medicine*, 27 (3): 577–593.

Baggerud (1959) [No title]. Letter dated 26th October 1959 from the Veterinary Directorate to NIPH. National Archive, Oslo: National Institute of Public Health, Virus Department, Box 1.

Clarke, A. (1987) "Research Materials and Reproductive Science in the United States, 1910–1940," in G.L. Geison (ed.), *Physiology in the American Context*. Bethesda, MD: American Physiological Society.

Clarke, A. and Fujimura, J. (1992) *The Right Tools for the Job? At Work in the Twentieth-Century Life Sciences*. Princeton, NJ: Princeton University Press.

Druglitrø, T. (2012) *Å skape en standard for velferd. Forsøksdyr i norsk biomedisin, 1953–1986*. PhD thesis. Oslo: Akademika.

Druglitrø, T. and Kirk, R. (2014) "Building Transnational Bodies: Norway and the International Development of Laboratory Animal Science, c.1956–1980," *Science in Context*, 27: 333–357.

Enders, J., Robbins, F. and Weller, T. (1954) "The Cultivation of the Poliomyelitis Viruses in Tissue Culture." Nobel Lecture, 11 December. Available at: www.nobelprize.org/nobel_prizes/medicine/laureates/1954/enders-robbins-weller-lecture.pdf.

Erichsen, S. (1963) "Kong Christian Den Femtes Norske Lov av 1687 og Dyreavdelingen," *SIFF Stalltidende*, 7. Oslo: National Library Archive.

Erichsen, S. (1964) "Stallinstruks nr. 17." National Archive, Oslo: National Institute of Public Health. Administration. Box 53.

Erichsen, S. (1965) [No title]. *Nytt fra Dyreavdelingen*, no. 10. Oslo: National Library Archive.

Erichsen, S. (1966) [No title]. *Nytt fra Dyreavdelingen*, no. 13. Oslo: National Library Archive.

Erichsen, S. (1967) [No title]. *Nytt fra Dyreavdelingen*, no. 16. Oslo: National Library Archive.

Erichsen, S. (1968a) [No title]. *Nytt fra Dyreavdelingen*, no. 18. Oslo: National Library Archive.

Erichsen, S. (1968b) "Avdelingsinstruks nr. 5: Bruk og rengjøring av burutstyr og vaskerom." National Archives, Oslo: National Institute of Public Health. Administration. Box. 53.

Erichsen, S. (1969) [No title]. *Nytt fra Dyreavdelingen*, no. 21. Oslo: National Library Archive.

Erichsen, S. (1970) [No title]. *Nytt fra Dyreavdelingen*, no. 24. Oslo: National Library Archive.

Erichsen, S. (1972) [No title]. *Nytt fra dyreavdelingen*, no. 28. Oslo: National Library Archive.

Erichsen, S. (1975) "Forsøksdyrene er nødvendige for vår livsform," *Forskningsnytt*, 2: 2–6.

Friese, C. (2013) "Realizing Potential in Translational Medicine: The UnCanny Emergence of Care as Science," *Current Anthropology*, 54 (S7): S129–S138.

Guerrini, A. (2003) *Experimenting with Humans and Animals: From Galen to Animal Rights*. Baltimore: Johns Hopkins University Press.

Haraway, D. (1989) *Primate Visions: Gender, Race, and Nature in the World of Modern Science*. New York: Routledge.

Haraway, D. (2008) *When Species Meet*. Minneapolis: Minnesota University Press.

Krag, C.H. (1960) [No title]. Letter from Krag to the NIPH, 23 April 1960. National Archives, Oslo: National Institute of Public Health. Virus Department. Box 1.

Lahelle, O. (1959a) [No title]. Letter to National Custom Clearance. 2 December 1959, National Archives, Oslo: National Institute of Public Health. Virus department. Box 1.

Lahelle, O. (1959b) [No title]. Letter dated 23 October 1959 to the Ministry of Agriculture, Directorate of Veterinary Medicine. National Archives, Oslo: National Institute of Public Health. Virus department. Box 1.

Lahelle, O. (1959c) [No title]. Note from Lahelle to Health Directorate, 13 October. National Archives, Oslo: National Institute of Public Health. Virus department. Box 1.

Law, J. and Lien, M. (forthcoming) "Practices of Fishy Sentience," in K. Asdal, T. Druglitrø, and S. Hinchliffe (eds), *Sentient Creatures: Humans, Animals and Biopolitics*. Farnham, UK: Ashgate.

Lewis, J., Atkinson, K., Harrington, J., and Featherstone, K. (2012) "Representation and Practical Accomplishment in the Laboratory: When is an Animal Model *Good-enough?*," *Sociology*, 47 (4): 776–792.

Mol, A., Moser, I., and Pols, J. (eds) (2010) *Care in Practice: Tinkering in Clinics, Homes and the Farm*. Amsterdam: Transkript verlag.

Oshinsky, D. (2010) "Polio," in A. Artenstein (ed.), *Vaccines: A Biography*. New York: Springer.

Rader, K. (2004) *Standardizing Animals for American Biomedical Research, 1900–1955*. Princeton, NJ: Princeton University Press.

Roth, W.T. (1969) "Supplying Wild Primates to the Laboratory," in *Using Primates in Medical Research. Part I: Husbandry and Technology*. New York: Karger. pp. 1–16.

Schmidt, L. (1972) "Problems and Opportunities of Breeding Primates," in WIB Beveridge (ed.), *Breeding Primates: Proceedings of the International Symposium on Breeding Non-Human Primates for Laboratory Use, Bern 1971*. New York: Karger, pp. 1–22.

Thompson, C. (2013) *Good Science. The Ethical Choreography of Stem Cell Research*. Cambridge, MA: MIT Press.

Timmermans, S. and Berg, M. (1997) “Standardization in Action: Achieving Local Universality through Medical Protocols,” *Social Studies of Science*, 27 (2): 273–305.

11 Care in the Cage

Materializing Moral Economies of Animal Care in the Biomedical Sciences, c.1945–

Robert G. W. Kirk

This chapter adopts a historical perspective to explore how material infrastructures have structured human–animal relationships within the biomedical "animal house," c.1945 to the present.[1] Following what has been characterized as the "ontological turn" in science studies (for example, Woolgar and Lezaun 2013), it is shown that material infrastructures serve to structure the multiple values that have informed and shaped human–animal relations within the experimental biomedical sciences (cf. Mol 2002; Law and Singleton 2005). By focusing on enactment as opposed to "knowledge" (or epistemology) alone, specific infrastructural objects (such as the animal cage) are revealed to be multiple (as opposed to singular objects). Such a perspective is one step toward empirically grounding Donna Haraway's contention that "the 'shared conditions of work' in an experimental lab[oratory] make us understand that entities with fully secured boundaries called possessive individuals (imagined as human or animal) are the wrong units for considering what is going on" (Haraway 2008, 70). Taking up this observation, and expanding it to include material infrastructure, reveals how multispecies sociocultural relations performatively shape and are shaped by the physical infrastructures that make up lived relations and shared labor in the animal house and laboratory. Collectively, enactments of "multispecies relations" are taken to form a dynamic, situated and emergent "moral economy" wherein the moral economy cannot be separated from factors that would properly be associated with a political economy of animal-dependent experimental science.

The concept of "moral economy" has been profitably imported into science studies (for example Kohler 1994; Strasser 2011) where it has been used to describe and contrast the systems of values that govern the exchange of knowledge and commodities within different scientific communities. Such usage, however, emphasizes the analysis of the values that shape legitimate exchanges within human cultures as opposed to exchanges and relations between human and nonhuman organisms. Thus Kohler charts how practices of exchanging *drosophila* mutants shaped a unique "moral economy" shaping the research culture that grew about this organism. But this usage has little to say about moral relations beyond the human. Kohler exclusively examined how values governing commodity exchange operated to shape human communities; the question of how researchers construed their moral relationship to the fly was left unasked. In contrast, Daston presents a different understanding of the moral economy, which is more concerned with exploring how notions of scientific objectivity "are not simply compatible with moral economies; they require moral economies" (Daston 1995, 3). Ways of knowing within science, contrary to common-sense distinctions, operate through economies of affects and emotions that are entirely compatible with and necessary to "rational" scientific practices such as quantification and standardization. Daston's "moral economy," therefore, appears compatible with a focus on performance and enactment. For Daston, a moral economy operates as a balancing system providing structure to the "web of affect-saturated values" which shape scientific work through virtue of being tied to specific activities that "anchor and entrench but do not determine it" (Daston 1995, 4). Within animal-dependent experimental sciences, the performance of animal care has increasingly formed the core of the moral economy, shaping, and being shaped by, the material infrastructure of the animal house. Which is to say moral concerns focused on the performance of care have been enacted through material infrastructure. Here, then, Daston's foregrounding of affects and emotions serves as a means to view the "moral economy" not as exclusively human but rather as a multispecies concern.

As such, this chapter explores how multiple and often conflicting values and concerns centered on animal care have taken form within specific moral economies, become materialized within physical infrastructures of the animal house, and thereby constituted new values and concerns. The performance of animal care, in this sense, is shown to have operated to constitute the object of care and has done so by materializing affect-saturated values within physical infrastructure. Such an argument draws on an emerging literature responding to Bruno Latour's invitation to seek the "missing mass of morality" (Latour 1992, 233). For Latour, "morality … like science and technology, is a heterogeneous institution constituted from a multiplicity of events, which depends at the same time on all modes of existence and in part … on the arrangement of technical apparatus" (Latour 2002, 254). Accordingly, Latour challenges the perception that technologies belong to the realm of means whilst morality addresses the question of ends. Within the biomedical sciences, the experimental use of animals raises a plethora of moral questions but answers rarely examine the extent to which morality is built in to the physical infrastructure and technologies of the animal house and laboratory. The laboratory animal cage, for instance, is not merely a means to an end. Rather, it can be understood to be a material instantiation of the conflicting and complex values that inform and shape the practice and ends of animal-dependent science. Accordingly, the physical infrastructure of laboratories and animals houses can be viewed as simultaneous means and ends, operating to materialize, stabilize and enact moral economies. Rather than purely functional, biomedical infrastructures such as the laboratory, animal houses, and animal cage, instantiate multiple and varied sociocultural, political, moral, economic, and scientific values, in forms suitable to secure locally determined ends. More specifically, the laboratory animal house and animal cage can be viewed as macro and micro physical instantiations of care practice. Just as the material form of Latour's desk drawer mechanism compels him to obey a moral principle, so too have animal cages embodied and expressed specific understandings of care. Understandings which, in turn, shape the possibilities of the relationships of multispecies concern that may or may not be shared within the animal house and laboratory.

This argument is developed through two complementary sections that explore how, why, and to what consequence a specific understanding of laboratory animal care, which prioritized animal health, was enacted through practices that came to be instantiated within the macro architectural infrastructure of the laboratory animal house (section 1) and the micro infrastructure of the animal cage (section 2). A final section discusses the implications of the historical entanglement of material infrastructures and care practices, relating this analysis of how moral economies of care have been enacted through material infrastructures to recent calls to explore new possibilities for acting with care within shared liveable worlds (Puig de la Bellacasa 2011).

Care in the Animal House: Institutionalizing the Moral Economy of Animal Care

Significant components of contemporary laboratory animal care practice emerged with and through twentieth-century historical trends that have been collectively characterized as "biomedicalization," being processes that place an intensive focus on the promotion of health through new biopolitical economies. Biomedicalization typically emphasizes the capacity of science and technology to transform bodies and produce new individuals and collectives (Clarke et al. 2010). Early twentieth-century laboratory animal care focused on health, specifically the need to prevent infectious disease (illustrating how biomedicalization operated across species boundaries). This is evident in the work of William Lane-Petter, an early proponent of the professionalization of laboratory animal care in the post Second World War period. Lane-Petter argued that

> the standard of laboratory animal accommodation to be aimed at is nearer to that of a hospital than of a farm, and terms like "farm" and "stable" should not be found in the vocabulary of this discipline [of animal care].
>
> (Lane-Petter 1959, 182)

In calling for a shift from the farm to the hospital, Lane-Petter was highlighting the specialist practices of breeding, husbandry, and care that he believed were necessary for the production of animals suitable for scientific use. As will become clear, it is not incidental that Lane-Petter communicated his vision of animal care through reference to physical architecture. Lane-Petter's choice of the hospital as the architectural model for the animal house derived from his principal definition of animal health and welfare in terms of freedom from infection and disease. The definition of animal health in terms of infection provides a paradigmatic example of how animal care enacted a moral economy that aligned animal welfare with animal health, scientific utility, fiscal costs and sociocultural values. Latent infection within laboratory animal stocks risked undermining experimental procedures. For instance, pathological manifestations of infection might be mistakenly thought to be a consequence of the experimental interventions resulting in the resultant knowledge being misleading and impossible to reproduce. Furthermore, infection could manifest as disease so rapidly that animals frequently failed to survive experiments in sufficient numbers to provide statistically meaningful results. At the same time, endemic infection had significant economic consequence as animals lost to disease were a wasted investment proving costly in research time and labor. The economic burden was intensified by the common practice of routinely using more animals than was statistically required to ensure enough lived to gain viable results (more animals meaning higher costs). Managing the problem of infection was particularly challenging as pathogens were often "latent," displaying no obvious visible signs until triggered by an environmental change to manifest as overt disease. In the absence of rigorous epidemiological knowledge, animal care was enacted through spatial and material techniques such as "quarantine" that were increasingly built into animal house and laboratory infrastructures.[2] As regimes of infection control became institutionalized within architectural design, animal care was gradually "materialized."

One consequence of the alignment of animal health to scientific utility was to raise the status of the animal house to the point that substantial investment in its infrastructure became desirable.[3] The Philadelphia-based Wistar Institute's rat house, for example, purpose built in 1922, enacted a moral economy of animal care through its physical infrastructure. The architecture of the building discouraged internal movement by providing an external entrance/exit to each room (making use of open air to minimize risk of cross-infection). This also allowed individual rooms to be effectively isolated should infections be discovered (Greenman and Duhring 1931).

By mid-century, animal care practice was inseparable from the wider infrastructure of the animal house as it was established practice for different animal stocks to be isolated in numerous small rooms to enhance health by minimizing cross-infection (Howie 1956). Writing in 1950, Leonell C. Strong of the Jackson Laboratory advised "it is extremely important to maintain a colony … by themselves and not in contact with any other species of animals … no new animals should be brought into the laboratory from any outside source" (Strong 1950, 80–81). Microbial loads and tolerances were inherited in the first instance and thereafter considered to be fragile ecologies which, in practice, determined which animal populations could be safely introduced and mixed and which could not. As pathogens such as "ring worm" were found to be zoonotic, the human body together with its microbial load became a key determinant of animal health, increasingly subjected to control through spatial and material distancing (albeit a distancing premised upon and emphasizing the shared microbial loads of human and animal). This approach reached a zenith with the development of isolator technologies to establish and maintain secure microbial environments; making possible the creation and maintenance of germ-free and specific-pathogen-free animals (Kirk 2012). These new techniques, developed and institutionalized through the 1950s, brought into being new forms of life with entirely defined microbial relations known variously as "clean," "pathogen free," "specific pathogen free," "disease free," and "gnotobiotic" (known life) animals. In contrast to conventional animals, these were presented as "sophisticated" having been "altered from, deprived of, primitive simplicity or naturalness" to become "the healthy animals we have been looking for for years" (Lane-Petter 1966, 54; Lane-Petter 1963, 55).

One of the earliest implementations of "pathogen free" breeding techniques in Britain was a purpose-built pharmaceutical research facility constructed for Imperial Chemical Industries Ltd. (ICI) on a 350-acre rural site at Alderley Park (Cheshire) in 1958. Having conducted a global survey of trends in biomedical institutions, ICI designed their entire site about principles of microbial control. Animal work was divided into three categories, spatially separated via the quarantine principle. Laboratories for "non-infectious" and "infectious" research were located at opposite ends of the main building linked by basic chemistry laboratories intended to act as a barrier between the two (Figure 11.1).

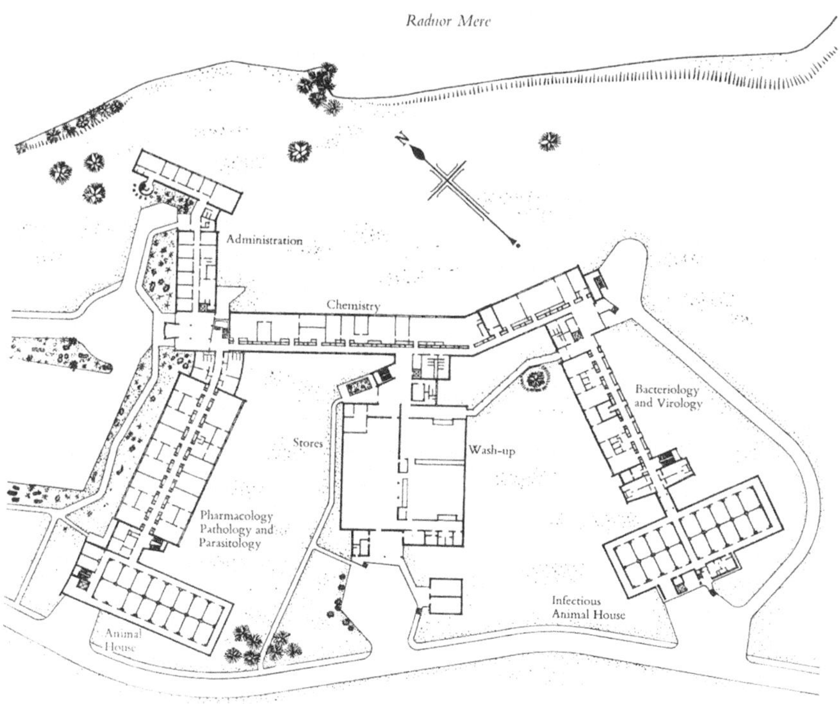

Figure 11.1 ICI main research facilities

Source: I.C.I., n.d.

Outputs from chemical and pharmaceutical research would flow down the east (non-infectious) or west (infectious) wing as appropriate. By removing the need for movement between the infectious and non-infectious research wings the building worked to minimize the risk of accidental transferal of microbes thereby promoting animal health. Within the building, internal passageways were minimized with entry to animal rooms restricted to outside doors. Floor-level physical barriers were incorporated to prevent escaped animals from roaming too far (Figure 11.2). Different animal rooms were color-coded to visibly alert workers to the fact they had moved from, for example, rooms where clean animals were being prepared for experimental use ("white") to laboratories where animals had been experimentally infected with pathogens dangerous to humans ("red"). In these ways, the building infrastructure actively shaped behavior enacting the moral economy of laboratory animal care.

Accordingly, the animal production unit was located some distance away within a 10-acre glade shielded by forestry with the surrounding lands protected from vermin and other wild animals by secure double fencing fitted with a double gated entry point (Figure 11.3).

The "Breeding Unit" was designed to maintain a highly controlled and microbially secure environment for the production of "Specific Pathogen Free" (SPF) animals. The building was physically divided into two areas, "dirty" and "clean," with all movement from one to the other regulated by mechanical barriers and rigorous decontamination controls. The windowless clean area had no "natural"

Figure 11.2 Entry to animal rooms was only possible from a perimeter open-air corridor (fine netting prevented wild animals and birds from gaining entry to the corridor). Strategically placed material barriers allowed human movement whilst preventing undesired rodent movements

Source: I.C.I., n.d.

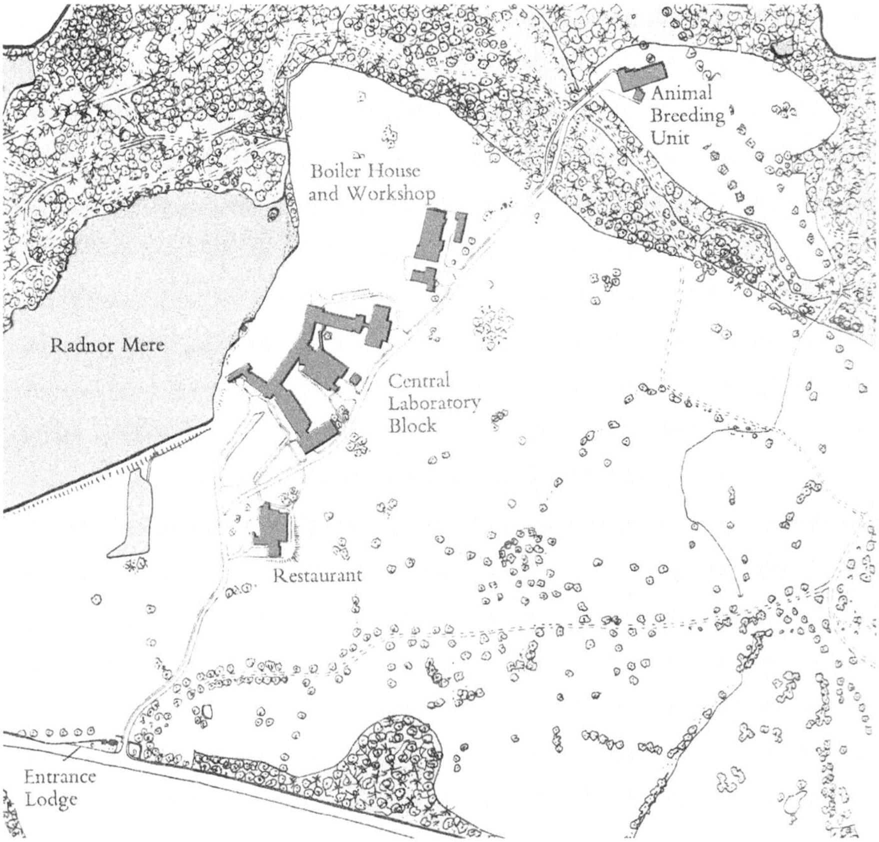

Figure 11.3 ICI site map illustrating spatial separation of SPF Animal Breeding Unit from research laboratories

Source: I.C.I., n.d.

source of ventilation or light. Air intake was artificially managed, drawn from a high altitude, and filtered, heated, conditioned and circulated throughout the building at a fixed rate of ten complete turn-overs per hour being additionally filtered on entrance/exit of each room. Material goods could only enter the clean area by passing through double-locked autoclaves designed so that the inner doors mechanically locked once an outer door was opened until an automated sterilizing cycle had completed. In this way, food and water were rigorously sterilized, with the latter delivered to animals via a purpose-designed automated watering system integrated into the cage racking. A similar system automatically flushed animal housing units with aseptic water washing away feces and other waste, again illustrating how the moral economy of care became integrated into the infrastructure of the building.

Human access to the clean area was also managed by infrastructural design (Davey 1959). Staff entered through a separate, smaller, outhouse where they were expected to strip and store all clothing and belongings before passing through a powerful "shower-bath" after which they entered a second room to dress in autoclaved sterile work-clothes. Entrance to the main building's clean area was through a closed linking corridor involving further decontamination procedures such as exposure to insecticides and ultra-violet light with each stage separated by an airlock (Figure 11.4). Nonetheless, excluded microbes soon took

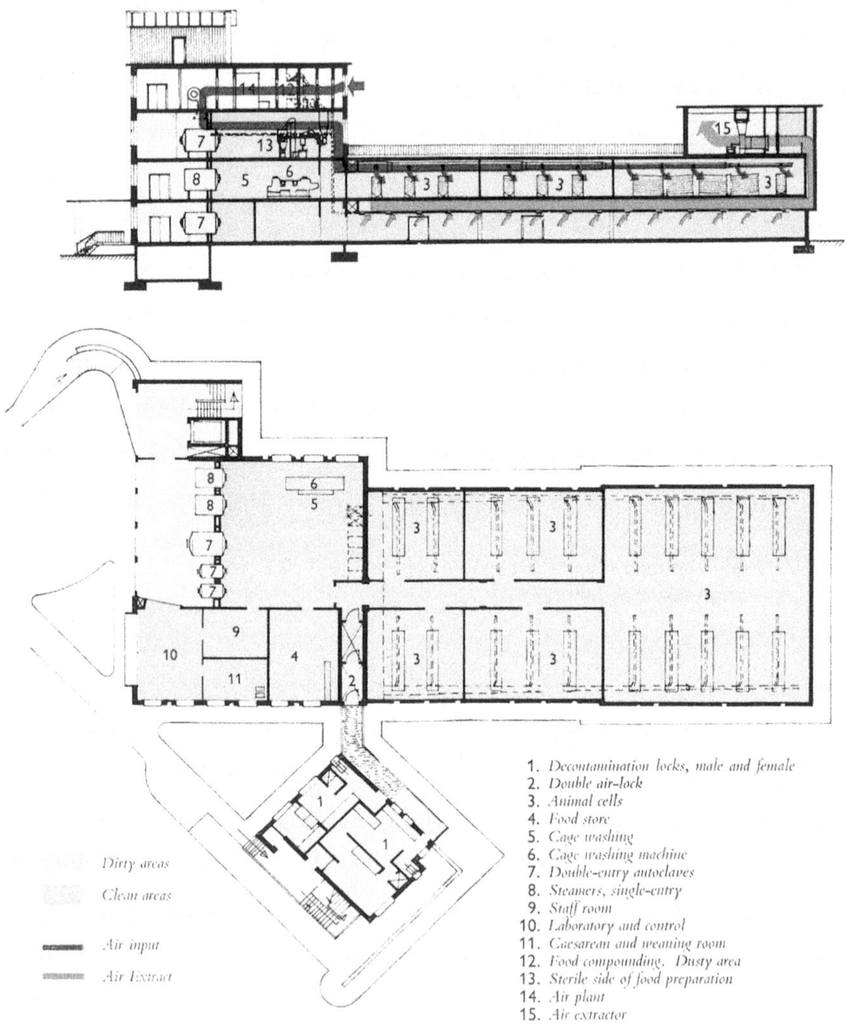

Figure 11.4 ICI Animal Breeding Unit

Source: I.C.I., n.d.

up residence in the clean area; one of the earliest to make its presence felt was *B. coli*. With time, SPF animals proved to be highly susceptible to infection with human microbes, with one explanation being that their artificially adjusted microbial flora lacked competing microbes. Human *Salmonella* carriers, for instance, were identified as a threat to SPF colonies which, once infected, subsequently served as vectors spreading the problem to human carers (cf. Hull 1963). Similarly, humans were identified as reservoirs for *Pseudomonas aeruginosa*, which devastated SPF murine populations (Flynn 1963; Hammond 1963). Consequently, new surveillance techniques were introduced to identify human carriers of pathogens so that they could be removed from duty (Van de Waaij et al. 1963). At the ICI breeding unit, routine checks quickly revealed that the SPF colonies had come to harbor proteus species, coliform organisms, and *Streptoccocus faecalis*. By 1961, *Pseudomonas pyocyaneus* and a human type of *Staphylococcus* had also been introduced to the shared microbial ecology presumably by human carriers. In general, such intrusions were tolerated as the microbes concerned were not considered to be pathogenic. Nevertheless, the ability of microbes to penetrate a building designed to be microbially secure illustrates how infrastructures intended to isolate and thus prevent microbial exchanges worked simultaneously to reveal the full extent to which human, animal and material infrastructures co-existed as a shared microbial ecology.

In spite of these problems, ICI researchers universally reported that the new SPF standard of animals were "fitter in every way" (Davey 1962, 8). Nor were they alone. These new practices created new forms of life guaranteed to be free of specific pathogens. Once initial stocks were introduced (via surgical removal of progeny before birth and raising by germ-free foster parents within the clean area) the breeding unit was capable of producing 100,000 rats and 500,000 mice annually. SPF animals were found to be healthier, exhibiting greater fertility and an extended life span in comparison with conventional animals. They were also thought to be more physically robust, better able to endure stressful experimental procedures and tolerating higher doses of toxins. SPF status, for example, made mice and rats plausible species for long-term experiments (such as the US Food and Drug Agency's demand that food additives be studied for two years in at least two species) which had hitherto, through necessity, employed larger more expensive species (Davey 1962). As such, the making of SPF animals illustrates how care practices gradually attuned forms of life to better meet scientific need. In a sense, SPF animals embodied the values of the moral economy in that they better met the needs of science, of fiscal economy and of the perceived needs of animal welfare. By the end of the 1960s, SPF criteria had become an accepted standard for the majority of commonly used laboratory species (Anonymous 1969).

The ICI complex at Alderley Park reveals how concerns over the health, welfare, scientific utility and fiscal cost of laboratory animals, which together formed a coherent moral economy of care, were enacted through the macro-infrastructure of the animal house and laboratory. So critical was the need to control microbial relations to the moral economy of laboratory life that it came to dominate the design of modern research facilities in the decades after the close of the Second World War. Traditional methods of quarantine (through spatial distribution and material barriers) were enhanced by new techniques, technologies and infrastructures, which together drove the creation of new microbially managed ecologies. SPF technologies instantiated "second natures" that controlled and enhanced the shared multispecies microbial ecology of lived relations within these new spaces. Accordingly, the decade and a half following the close of the Second World War witnessed a radical shift in the meaning and ontology of what it was to be a laboratory animal. By the end of the 1960s, laboratory animals were precisely what their name suggested: new forms of life that were born of and for the laboratory. By giving material form to a specific moral economy, the macro-infrastructure of animal houses and laboratories literally institutionalized care practices that subsequently transformed the laboratory animal into a new form of life defined by their microbial loads and shared ecological relations.

Care in the Cage: Materializing the Moral Economy of the Animal House

Not all, however, possessed the capital for building the moral economy of animal care into the macro-infrastructure of animal house and laboratory architecture as did ICI. Such projects in general worked best when creating new facilities as opposed to adapting established buildings and/or animal stocks (the major obstacle being the latter possessed complex microbial ecologies that were difficult to erase). Consequently, alternative approaches to enacting moral economies of laboratory animal care were developed focused on micro-infrastructure such as the animal cage. In 1963, for instance, the University of Cambridge's mouse house developed an innovative cage design making use of a new plastic. Somewhat controversially, the plastic chosen could not withstand the heat of autoclaving. Describing the new cage, Margaret Wallace acknowledged the "hesitation by other laboratories" to use such plastics yet proposed nonetheless that it was "the cheapest plastic, least noisy [and] sterilization is usually unnecessary if hygiene is studied in cage design and all parts of the laboratory" (Wallace 1963, 69). Prior to the establishment of SPF as an expected health standard, many held the view that "a completely disease-free stock of laboratory animals is an impossible ideal" (Salaman 1956). From such a perspective, animal care was envisioned as a continuous and responsive activity, a practice that called for the dynamic management of the often conflicting needs that made up the moral economy of care (which variously included animal health and welfare, human labor, experimental requirements and economic considerations). Thus, it was perfectly acceptable to weaken one element of this constellation in favor of strengthening others that were of local importance. Whilst SPF techniques removed the risk of infection, almost entirely unless mistakes were made, it did so at a financial and labor cost that many, at least initially, felt unnecessary and/or impractical. Wallace described the five values that had shaped Cambridge's particular instantiation of the moral economy of care: (1) maximize animal health and breeding rates, (2) minimum demand on human labor, (3) materials of cage maximize 1 and 2 at the minimum of economic cost, (4) promote experimental utility, (5) materials to be conducive to a pleasant environment for the worker. Whilst some of these values could be productively balanced, rarely did progress in one area not have unforeseen consequences elsewhere. Materializing these values within the micro-infrastructure of animal houses and laboratories required calculated compromise. As different laboratories favored different ways of balancing the moral economy to suit local purposes, there was general agreement of "an urgent need for some experienced body to undertake either basic research on behavior of caged animals and the testing of designs" as "only then will standardization become possible" (Wallace 1963, 65). Wallace was not the first, nor the last to attach this caveat to their description of local practices. In 1949, for instance, a review of laboratory animal housing practice observed that "most workers express a preference for standard sizes but in regard to these same standard sizes, it appears that *quot homines, tot sententiae*" (Jones and Wood 1949, 194). There were as many standards as persons.

One obstacle to the development of shared standards and systematic knowledge was that animals responded unpredictably to novel circumstances and even the most controlled of environments contained a constellation of factors that could impact on an animal's make up. Even the smallest of events, such as the appearance of a stranger in the animal house, could detrimentally impact on animal welfare (recognized principally at the time through depressed reproduction rates). Dramatic change, such as moving an animal population from one room to another, had been known to prevent normal breeding for several months (Farris 1945, 4). Assessing the consequences of different technologies and practices of care was therefore far from straightforward. In the 1950s, for instance, there was a general trend to replace wooden cages with metal and wire mesh cages as the latter could be autoclaved and thus were presented as a hygienic intervention to deter cross-infection and promote health. Metal, unlike wood, was less likely to harbor parasites and pathogens. However, some found that certain species, particularly mice, were less productive in metal cages. One explanation for this was that mice preferred "darkness" resulting in recommendations for the continued use of traditional wooden boxes (Lane-Petter 1959, 28). Others attributed the problem to the colder environment, claiming that mice would reproduce normally if the overall temperature of the room was raised (Strong 1950, 84). Hence *quot homines, tot sententiae*; scope for interpretive flexibility ensured that were as many opinions as persons. Nevertheless, as the cage was gradually established as a core means to enact animal care, systematic knowledge of its properties and impact on animal health and welfare became increasingly important:

> To keep an animal in a cage will inevitably modify its behaviour in several ways, not always to the disadvantage of the animal. Certain types or sizes of cage may affect the animal adversely, and offend both humanitarian considerations and also scientific requirements … An unnecessarily large cage will

> take up so much room in the animal house that the already high cost of maintenance will be fruitlessly multiplied. A cage of dimensions smaller than those dictated by custom, conjecture, anthropomorphic misconception … may not necessarily interfere with the well-being of its inmate. It is quite unjustifiable to assume that the bigger the cage, the better for the animal … There is a great need for some real information on the subject. Cages are expensive items of equipment and there is such a multiplicity of design that standardization remains unattainable in the absence of precise knowledge: Yet standardization of, say mouse boxes, would reduce the cost considerably.
> (Lane-Petter 1953, 126–127)

Such calls reflected growing recognition of the difficulty of holding in productive tension the varied and often conflicting values that shaped moral economies of care as much as the challenge of enacting moral economies through material infrastructure. Custom, conjecture, and anthropomorphic misconceptions were no longer acceptable criteria for determining how these competing values should take material form in the cage. Instead, animals and their environments would become the *object* of study in themselves, objects studied, made and remade largely through practice. As a result, the cage ceased (if it ever had been) to serve merely as a means to an end, which is to say an epistemologically and morally empty object. Instead, the cage increasingly became a material embodiment of a range of moral, ethical, economic and epistemological concerns that together made up a moral economy of care.

Responses to the rodent preference for coprophagy demonstrate the way cages enacted moral economies of care. Separating animals from their feces appeared an obvious and necessary step toward establishing a hygienic and thus healthy animal house environment. One simple and effective means to achieve this was the introduction of cages with wire mesh flooring. This allowed feces to fall into a lower tray out of reach of the animal to be easily cleaned by the animal caretaker (or automatically flushed as at ICI). However, rodents were known to consume their own feces and it was unclear what if any impact preventing this practice might have on their health, welfare, reproductive efficiency and scientific utility. Understanding coprophagy's role was far from straightforward as various additional factors – such as the nutritional content of food, which varied across locations despite *numerous* standard formulations emerging – complicated coherent assessments. Further difficulties emerged in understanding the social role of coprophagy, as some rodents were known to eat feces from another's behind if housed in groups that allowed it. Within specific fields, such as nutritional research, coprophagy gained enhanced importance as it impacted directly on the object of research. Coprophagy introduced an uncontrolled source of nutrients, whilst its prevention risked working with an animal that departed from the norm in unspecified ways. Moreover, nutritional research's experimental designs routinely incorporated the removal and analysis of feces as a standard practice. Consequently, coprophagy spawned a lively research culture enacted through new forms of caging. Numerous variations of what came to be known as "anticoprophagy" or "metabolism" cages were developed to separate rodents from feces. One popular design consisted of narrow wire-mesh cylinders connected to form a simple circular tunnel. This allowed rodents to crawl forward and obtain food but made coprophagy impossible by preventing turning within the tube (Chalam Metta et al. 1961, 474). The tubular coprophagy cage favored the needs of experimental science over the health and comfort of the animal as prolonged confinement restricted a range of "normal" behaviors (Lane-Petter 1957, 30). An alternative, designed by George Brownlee and used at the Wellcome Physiological Research Laboratories (UK), consisted of a cubic metal cage with a cylinder at the end of which a feeding cup was located. Rodents could enter the cylinder to feed but could not turn about within the cylinder, thus feces dropped through the wire mesh floor with little opportunity for consumption. This design attempted a better balance between animal health, welfare and experimental necessity by separating rodent from feces without overly restricting movement (Gorer 1947, 156). The distinction between the two lay not so much in the concept or aim but in the specific choices over how to enact the local moral economy through the materiality of the cage. As researchers looked first to local need, a multiplicity of cage designs emerged sustaining the role of custom, conjecture, and anthropomorphic misconceptions to shape the material instantiation of moral economies even as the work was conducted with the intent of systematically establishing shared standards of practice.

Concerns over coprophagy declined rapidly with the widespread adoption of SPF standards as the controlled microbial ecology of an SPF facility removed any health threat. However, the eventual acceptance of SPF animals as an expected standard in large part became possible once the technique could be enacted at the micro-infrastructural level of the cage. Work to this end began in the late 1950s with the recognition that traditional open-top cages, whilst efficient in allowing easy access to animals, also encouraged the transmission of pathogens. One of the first so-called "filter" cages was developed by Lisbeth M. Kraft at Yale for the purpose of investigating and eradicating a local endemic infection causing diarrhea in infant mice (Kraft 1958). Kraft's system used fiberglass mesh to establish a secure barrier preventing the free exchange of air between the cage and wider environment of the animal house (Figure 11.5). Accordingly, individual filter cages could only be opened within a large isolator fitted with its own clean air supply.

Through the 1960s, Kraft's filter cage was adapted to a myriad of new applications as it provided an affordable means to remove pathogens from animal houses without the economic costs required by macro-infrastructural SPF facilities such as that of ICI (e.g. Simmons et al. 1967). Nevertheless, whilst filter cages provided an effective means to enact a moral economy that prioritized freedom from infection it came at the cost of a basic animal care practice: Routine observation. Sealed in metal cylinders, animals could only be viewed by examination within a secure isolator – a labor-intensive activity that risked additionally stressing rodents. It was not until the advent of robust and affordable plastics in the 1970s that a better balance could be struck between the pursuit of health and the observational needs of animal care.

Figure 11.5 Original "Filter" cages designed by Kraft. The cage on the left is shown without the fiberglass filter lid; that on the right is the complete unit

Source: Kraft, 1958.

Attempts to make productive use of plastic within cage design had been made since at least the 1940s. However, plastic was initially prohibitively expensive and unable to compete with metal which, unlike early plastics, could withstand the high heat of the autoclave (Strong 1950, 86–87). Though the 1960s advances in plastic production produced first polycarbonate and later polysulfone, making possible affordable, lighter, warmer, and importantly transparent caging that allowed the observation of animals without their physical disturbance. By the mid-1970s, filtered roof techniques began to be combined with newly emerging plastic cages to produce an effective and affordable microbially secure micro-environment. One of the first widely adopted commercial examples was that of Robert Sedlacek (Sedlacek and Mason 1977) marketed as Lab Products Inc.'s "Micro-Isolator™ System." Within the "Micro-Isolator™ System" mice existed in a secure microbial environment separated from that of the wider animal facility. The adoption of isolation cages provided a simple material solution to previously complex ecological and practical relations. The SPF conditions of Micro-Isolator™ caging, for instance, rendered feces effectively "clean" – rendering coprophagy a non-problem.

Nevertheless, as with previous innovations, new material infrastructures produced new challenges to animal care demanding further innovation. Isolation cages effectively created microclimates that increased moisture accumulation and thus temperature relative to the room as a whole. Again, the attempt to create independent ecological environments served only to re-inscribe their co-dependence as increased temperatures within isolation cages necessitated lower temperatures within the wider animal house. Care practice continued to require the management of the whole. A more significant problem caused by the micro-climate effect was the tendency for ammonia levels to detrimentally rise to the point at which mice were gradually suffocated by their own urine (for example Keller et al. 1989). Here again, local contingencies shaped experience. When developing the Micro-Isolator™ System, Sedlacek had not encountered this problem as he had worked with "germ-free" mice that lacked the urease bacteria required to convert urine into ammonia. Once more the complexity of microbial relations within the laboratory ecologies came to the fore as urine replaced feces as the primary health concern. In contemporary mouse facilities, this problem has been reduced by the development of Individually Ventilated Cage (IVC) systems where each isolation cage has its own regularly circulated air supply. Multiple IVC cages had emerged in the 1960s in response to various local needs. One early IVC was designed to allow work on dangerous pathogens to be pursued relatively safely by safeguarding researchers, technicians and animal carers from accidental exposure to infected animals (Cook 1968). Others, such as an IVC designed by Edwin P. Les of the Jackson Laboratory, Bar Harbor (Maine, USA) and Bill Thomas of Thoren Industries, was intended as a means to promote animal health and welfare and thus materialized a moral economy of animal care.[4] As such, the IVC again illustrates the complexity of concerns and values that were enacted in the material form of the animal cage.

Conclusion: The Materialization of Care

The "cleaning up" of animals has been a prominent theme of this chapter. On her journey from cyborg to companion species concerns, Haraway paused to offer a twenty-one point program within which a "key question is who cleans up the shit in companion species relations?" (Haraway 2003, 79). We may accordingly propose that "cleaning up the shit" is an act of care. Whether by automatic systems integrated into the architecture of animal house buildings, such as the mechanical flushing away of feces and waste from animal cages within the SPF facilities at Alderley Park, or the physical separation of animal from feces using wire-mesh floored cages, removing shit was a critical concern for those who worked with and cared for laboratory animals. It was a concern that formed a central part of moral economies of animal care which, in various ways, came to be embedded in and practiced through the macro and micro material infrastructures of animal houses and laboratories. Cleaning up the shit was neither a simple practice nor was it without transformative consequence. Shit was host to a multitude of microbial species all of which had to be understood and managed through their relations with the wider laboratory ecology, human, animal and material alike. Choices had to be made as to which microbes were to be included or excluded from the shared ecologies of laboratory life, choices that balanced the competing concerns and values which shaped local moral economies of care. In some instances, competition for shit posed significant challenges. When feces became a matter of scientific interest, researchers laid claim to a resource that rodents utilized to meet little understood nutritional and possibly social needs. Competing demands were held in a productive tension by enacting these concerns through material infrastructures, stabilizing (for a time) animal care practices. As each new cage embodied and expressed a specific moral economy of care it also generated new challenges driving innovation and historical change.

Over time, the material infrastructures of the animal house and laboratory have been repeatedly configured and reconfigured in response to shifting moral economies of care. The changing material infrastructures of the laboratory animal house can be understood as having co-evolved with new forms of life (e.g. SPF animals), new animal care practices, new human identities (e.g. professional animal caretaker), research trajectories, and other factors that, cumulatively, made up a dynamic moral economy of care. Material infrastructures have facilitated the enactment of relations between forms of life that were recognized to possess insecure boundaries. From this perspective, the laboratory and animal house become lively relational sites where the process of becoming anew is always a potential. Material infrastructure can be seen as collaborative partners instantiating and shaping changing moral economies of laboratory life and labor. Material infrastructures, in sum, serve to intervene in processes of shared becoming. Understanding such processes, which would be to cultivate a reflective awareness of the multiple ways of practicing care through material infrastructures, would form one response to recent calls to theorize care as a "living technology with vital material implications for human and non-human worlds" (Puig de la Bellacasa 2011, 101).

Notes

1 Importantly, the focus of analysis is less the "laboratory" than the "animal house." This is an important distinction, historically, albeit one that has become less so in recent years as the boundary between the two has been eroded due to experimental procedures increasingly moving from a distinct laboratory space to what is now commonly known as the "Biological Services Facility."

2 Early studies of the epidemiology of laboratory animals had reported little more than the complexity of the problem (e.g. Greenwood et al. 1936). Without precise epidemiological knowledge, it was difficult to meaningfully attribute a disease outbreak to new microbes having entered the microbial ecology of the animal house or whether the introduction of new animals had initiated a process of social or environmental change that triggered an already present latent infection to manifest as disease.

3 Nevertheless, the animal house continued to be secondary to the laboratory. Writing in 1959, William Lane-Petter complained of "the bitter experience of many that this consideration has all too often been overlooked. In the planning, any surplus on the overall budget has been grudgingly given over to animals, which have suffered on the

way by any economies that are forced upon the planners. The wise director takes his 10 or 15 percent at the beginning and builds his animal house before the laboratory. He can then establish it while the laboratory is being completed, so that his staff can move in when the painters move out and have animals ready to hand" (Lane-Petter 1959, 193).

4 Marketed as the "Maxi-Miser® Positive Individually Ventilated System (PIV)" from 1978. Thoren Industries was the first commercial company to develop warm surface plastic cages in 1953.

References

Anonymous (1969) "Animal for research," *Lancet*, 7620: 582.

Chalam Metta, V., Nash, L., and Connor Johnson, B. (1961) "A tubular coprophagy cage for the rat. *Journal of Nutrition* 74: 473–476.

Clarke, A., Mamo, L., and Fosket, J.R. (2010) *Biomedicalization: Technoscience, Health, and Illness in the U.S.* Durham, NC: Duke University Press.

Cook, R.O. (1968) "New Ventilated Isolation Cage," *Applied Microbiology*, 16: 762–771.

Daston, L. (1995) "The Moral Economy of Science," *Osiris*, 10: 2–24.

Davey, D.G. (1959) "Establishing and Maintaining a Colony of Specific Pathogen Free Mice, Rats and Guinea-Pigs," *Quality in Laboratory Animals (Collected Papers Laboratory Animals Bureau 8)*. London: MRC, pp. 17–34.

Davey, D.G. (1962) "The Provision and Use of Pathogen-free Laboratory Animals," *Proceedings of the Royal Society of Medicine*, 55: 253–263.

Farris, E.J. (1945) "Introduction to the conference on animal colony maintenance," *Annals of the New York Academy of Sciences*, 46: 3–4.

Flynn, R.J. (1963) "*Pseudomonas Aeruginosa* Infection and Its Effects on Biological and Medical Research," *Laboratory Animal Care*, 13: 1–6.

Gorer, P.A. (1947) "The Mouse," in A.N. Worden (ed.), *The UFAW Handbook on the Care and Management of Laboratory Animals*. London: Bailliere, Tindall and Cox, pp. 150–167.

Greenman, M.J., and Duhring, F.L. (1931) *Breeding and Care of the Albino Rat*. Philadelphia: Wistar Institute.

Greenwood, M., Hill, A.B., Topley, W.W.C., and Wilson, J. (1936) *Experimental Epidemiology: Special Report Medical Research Council no. 209*. London, HMSO.

Hammond, C.W. (1963) "*Pseudomonas Aeruginosa* Infection and Its Effects on Radiobiological Research," *Laboratory Animal Care*, 13: 6–11.

Haraway, D.J. (2003) "Cyborgs to Companion Species: Reconfiguring Kin in Technoscience," in D. Ihde and E. Selinger (eds), *Chasing Technoscience*. Bloomington: Indiana University Press, pp. 58–82.

Haraway, D.J. (2008) *When Species Meet*. Minneapolis: University of Minnesota Press.

Howie, J.W. (1956) "Chairman's Remarks," in *Infections in Laboratory Animals (Collected Papers Laboratory Animals Bureau 4)*. London: MRC, pp. 5–6.

Hull, T.G. (1963) *Diseases Transmitted from Animals to Man* (5th edn). Springfield, IL: Charles C. Thomas.

I.C.I. (n.d.) *Pharmaceutical Research*. Birmingham, UK: Kynoch Press.

Jones, J.I.M., and Wood, E.C. (1949) "The Housing of Laboratory Animals," *Journal of Hygiene*, 47: 190–196.

Keller, L.S.F., White, W.J., Snider, M.T., and Lang, C.M. (1989) "An Evaluation of Intra-

Cage Ventilation in Three Animal Caging Systems," *Laboratory Animal Science*, 39: 237–242.
Kirk, R.G.W. (2012) "'Standardization through Mechanization': Germ-Free Life and the Engineering of the Ideal Laboratory Animal," *Technology and Culture*, 53: 61–93.
Kohler, R.E. (1994) *Lords of the Fly: Drosophila Genetics and the Experimental Life* Chicago: University of Chicago Press.
Kraft, L.M. (1958) "Observations on the Control and Natural History of Epidemic Diarrhea of Infant Mice," *Yale Journal of Biology and Medicine*, 31: 121–137.
Lane-Petter, W. (1953) "Some Behavioural Problems in Common Laboratory Animals," *BJAB*, 1: 124–127.
Lane-Petter, W. (1957) "Animal House Equipment," in A.N. Worden and W. Lane-Petter (eds), *The UFAW Handbook on the Care and Management of Laboratory Animals.* London: UFAW, pp. 23–57.
Lane-Petter, W. (1959) "The Place of Laboratory Animals in the Scientific Life of a Country," *Impact of Science on Society*, 9: 178–196.
Lane-Petter, W. (1963) "Discussion," *The Choice of Experimental Animal: Laboratory Animals Centre Collected Papers (vol. 12)*. London: MRC, pp. 54–55.
Lane-Petter, W. (1966) "Sophisticated Laboratory Animals," in I. Gilliland and J. Francis (eds), *The Scientific Basis of Medicine: Annual* Reviews. London: Athlone, pp. 54–70.
Latour, B. (1992) "Where are the Missing Masses? Sociology of a Few Mundane Artefacts," in W. Bijker and J. Law (eds), *Shaping Technology – Building Society: Studies in Sociotechnical Change*. Cambridge, MA: MIT Press, pp. 225–229.
Latour, B. (2002) "Morality and Technology The End of the Means," *Theory, Culture and Society*, 19: 247–260.
Law, J., and Singleton, V. (2005) "Object Lessons," *Organization*, 12: 331–355.
Mol, A. (2002) *The Body Multiple: Ontology in Medical Practice.* Durham, NC: Duke University Press.
Puig de la Bellacasa, M. (2011) "Matters of Care in Technoscience: Assembling Neglected Things," *Social Studies of Science*, 41: 85–106.
Salaman, M.H. (1956) "Discussion," in *Infections in Laboratory Animals (Collected Papers Laboratory Animals Bureau 4)*. London: MRC, p. 17.
Sedlacek, R.S., and Mason, K.A. (1977) "A Simple and Inexpensive Method for Maintaining a Defined Flora Mouse Colony," *Laboratory Animal Science*, 27: 667–670.
Simmons, M.L., Richter, C.B., Franklin, J., and Tennant, R.W. (1967) "Prevention of Infectious Diseases in Experimental Mice," *Proceedings of the Society for Experimental Biology and Medicine*, 126: 830–837.
Strasser, B.J. (2011) "The Experimenter's Museum: GenBank, Natural History, and the Moral Economies of Biomedicine," *Isis*, 102: 60–96.
Strong, L.C. (1950) "The Care of Experimental Mice," in E. J. Farris (ed.) *The Care and Breeding of Laboratory Animals*. New York: John Wiley.
Wallace, M.E. (1963) "Cage Design Principles, Practice and Cost," *Journal of the Animal Technicians Association*, 14: 65–72.
Woolgar, S., and Lezaun, J. (2013) "The Wrong Bin Bag: A Turn to Ontology in Science and Technology Studies," *Social Studies of Science*, 43: 321–340.
Van de Waaij, D., Zimmerman, W.M.T., and van Bekkum, D.W. (1963) "An Outbreak of *Pseudomonas Aeruginosa* Infection in a Colony Previously Free of This Infection," *Laboratory Animal Care*, 13: 46–53.

12 The Spatial Arrangements of Making Research Piglets into Resources for Translational Medicine

Mette N. Svendsen

The creation of new medical knowledge has for centuries relied on animal experimentation in which the animal is made into a model of man. Within recent decades, animal models have not only been seen as a means to gain insights into human biology, but have also been pictured as central resources in creating closer connections between laboratory research and health-care services, an endeavor that is often termed "translational medicine." In the rhetoric of translational medicine, the use of animal models plays a critical role in testing drugs, medical procedures or nutritional regimes before introducing them in the clinic. As Lesley Sharp (2013) points out in her ethnography on experimental transplant medicine, there is nothing self-evident about resource-imagination and resource-making. To imagine something as a resource and to make it into a legitimate and available resource takes work, involves claims to specific pasts and futures and carries moral reasoning (Ferry and Limbert 2008; Sharp 2013). Following this, I propose that a consideration of the spatiality at stake in establishing research animals as resources for translational medicine will make us alert to what it takes morally and practically to connect animals to human health.

In this chapter I explore how piglets are housed, cared for and processed to become a resource for translational medicine. I move into the field of neonatology in Denmark in which piglets are born prematurely to model the human infant in the neonatal intensive care unit (NICU). I investigate the spatial arrangements in three different sites in the infrastructure of translational medicine: the pig laboratory in which piglets are born, taken care of and in the end killed; the semi-public spaces of scientific conferences in which pig-based research is presented to various actors in the field of translational research; the NICU where premature or seriously ill infants are treated – the infants whose health is seen as the ultimate goal of the life, suffering and death of the research piglets.

My concept of spatial arrangement includes a consideration of *the organization of space* around animal and infant (what is the character of their space, who and what is next to them, where are they allowed to move, what can be known about them in specific spaces) and *imaginations and expectations about the spaces* the human and the animal will inhabit in their lives (to enter the dissection lab, to enter a home). In applying this symmetrical approach to the piglet in the lab as well as the infant in the NICU, I do not in advance determine who is considered worthwhile for experimentation and who is worthy of a space in the human NICU, but I investigate how a being (piglet as well as infant) becomes configured as worthy of specific involvements in specific spaces. As such my study belongs to an expanding field of multispecies ethnography (Kirksey and Helmreich 2010) that explores the dynamic ways in which humans and nonhumans engage in meaningful relationships (Candea 2010; Kohn 2007) and the processes through which animal lives and deaths are intertwined with the making of human biology and possibilities for (human) ways of living and dying (Friese 2013; Franklin 2007; Sharp 2013).

I begin by introducing my interest in the spatiality at stake in making piglets available as resources for science, and my use of the concepts of *zoe* and *bios*. Following this I provide a short outline of the role of pigs in the Danish agricultural and biomedical landscape and introduce my case study. I then turn to the analysis and explore the spatial arrangements in the infrastructure of translational medicine focusing on daily practices in the pig laboratory, scientific presentations of the pig studies, and practices around infants in the NICU. Based on ethnography from these three sites I argue that zones of indistinction as well as distinct boundaries between animal and human characterize the way the piglets become resources for translational medicine.

The Spatial Constitution of Life

My starting point in the following is that the organization of space for humans involves the organization of space for animals. Reviel Netz's thought-provoking book *Barbed Wire* (Netz 2004) is helpful in this respect. Netz follows the invention and spread of the technology of barbed wire in America from 1874 to 1954 and provides a compelling perspective on physical control over space, highlighting how the prevention of movement through enclosures tells a history of humans, animals, and the physical world inextricably tied together. This perspective is also prominent in writings on the spatial dimensions of human–animal encounters (Buller 2013; Davies 2000; Jones 2000) that direct attention to the ways life is organized in confining certain humans and animals out in a field, inside a house, or away from cities or specific institutions. To be "in place" or "out of place" often implies very different human/nonhuman relationships (Jones 2000, 270). Situating my study in this perspective, I explore the spatiality at stake in letting the piglets pave the much-praised road between laboratory and clinic, which we may conceptualize as a process of making the animals *bioavailable*. In pharmacology, bioavailability is a well-known concept that refers to the degree to which a substance can be utilized by those parts of the body on which it is intended to have an effect. Here, however, I follow Aditya Bharadwaj (2008) in letting the concept refer also to the geopolitical, moral and social processes through which a body becomes ready to enter specific scientific or clinical practices.

In exploring how piglets are made bioavailable I have found inspiration in Giorgio Agamben's concept of "bare life" (Agamben 1998), which rests upon the distinction between *zoe* and *bios*. Agamben describes *zoe* as the mere fact of being alive, the bare life that is common to all living beings, whereas *bios* refers to a qualified life, a particular way of life within a human society (Agamben 1998). The distinction between *zoe* and *bios* reflects other conceptual pairs such as the distinction between *life* and *living* (Marsland and Prince 2012; Wahlberg 2008) and the distinction between *biological life* and *biographical life* (Svendsen 2011; Waldby 2002). In the following I use the concept of *zoe* to refer to life as living objects and biological instruments and I use the concept of *bios* to refer to sentient individual life that has subjectivity and is placed in affective relations within a sociality. Importantly, I do not treat *zoe* and *bios* as ontological states of being (cf. Marsland and Prince 2012). Rather, I set out to understand the socio-spatial arrangements that in specific situations produce some beings as *zoe* and others as *bios*. My point of departure is that actors – humans and nonhumans – are always part of a larger constellation and that we need to attend to how beings are enacted and allowed to act within this constellation (Mol 2013). What becomes a qualified biographical life and what becomes a biological bare life is neither inherent in bodies and brains nor socially determined, but an effect of socio-material practices. In particular, I trace the spatiality at stake in balancing the piglets as *zoe* and *bios* when making them bioavailable in the field of translational medicine.

The Study[1]

The fieldwork presented in this chapter comes out of a four-year research project starting in 2009 and exploring boundaries and relationships between human and animal in biomedicine. As part of this study I have followed experimental practices on research piglets in a perinatal pig laboratory in Denmark through shorter periods of observing laboratory practices and participating in research events and seminars in which laboratory results were discussed. In addition, I have conducted one month of almost daily field observations in the NICU in which I followed clinicians in their daily work of caring for preterm infants or seriously ill term infants. Since 2012 I have collaborated closely with three graduate students who also carry out observations and interviews in laboratory and NICU settings. We have conducted interviews together and share fieldwork data. My movements back and forth between the perinatal pig laboratory and the NICU made me alert to the practices that simultaneously divide and connect the spaces in which piglets and humans are allowed to act. In exploring these practices, my analysis draws on ethnographic observations in the perinatal pig laboratory in 2009, 2011, and 2013, and in the NICU at the university hospital in Copenhagen in 2010.

Pigs in Danish Biomedicine

For centuries, pigs have been an economic and national resource for Denmark. Breeding pigs for food has been based on systematic selection and has resulted in famous Danish pigs with leaner meat and an extra rib. At present, the Danish population is less than six million people whereas the number of pigs produced per year amounts to approximately 25 million. In Denmark, pigs are creatures that are not considered exotic and unpredictable others, but mundane production animals that contribute to a familiar fabric of living. Pork is a staple diet of most Danes; it is in the lunch packages and on the supper table several times a week in most families.

Despite its great volume, pig production is not particularly visible. Driving through the countryside in Denmark, the passer-by will probably not see one single pig, and only insiders may recognize the presence of the pig when passing slurry tanks next to low-rise animal houses and then be reminded of the country's export of high-quality pork meat, as well as of continuous public discussions of the pollution of Danish waters due to discharge of slurry. On a sunny day from the window of the airplane going from one end of the country to the other, one will notice that most of the land is divided into cultivated fields, yet few will realize that what is cultivated in a great number of the fields is fodder for pigs and other farm animals. Despite their public invisibility the pig shapes the landscape as well as the welfare of the country as agribusiness contributes a sizable portion of the Danish state's budget, thereby representing a revenue through which the welfare state funds itself.

Since the opening of the Royal Veterinary and Agricultural University in Copenhagen in 1858, the meat-producing industry has had close relations with Danish veterinary science. Within recent decades, veterinary science has increasingly moved from studying animal health to focusing on human health via animals. For example, the research I follow places the piglet and the infant on the same biological scale and considers the piglet a tool for creating knowledge about an inflammatory bowel disease called necrotizing enterocolitis (NEC), one of the most common and devastating diseases in premature infants (Neu and Walker 2011). Danish production pigs happen to be excellent models of NEC, which might be related to the fact that the large litters of Danish production sows counteract organ maturation in newborn piglets and hence make them more sensitive to NEC than other races of pigs (Sangild et al. 2013). Also, international recognition of important phenotypic as well as genotypic traits between humans and pigs has supported the use of pigs as models of the human organism in biomedical research (Groenen 2012; Kuzmuk and Schook 2011).

In Denmark the use of pigs in biomedicine has not provoked public disapproval. In the Danish cultural landscape the pig is as commonplace as the soil Danes live on; an understanding of the pig that we may see as intimately connected to the disciplining logic of extensive farming in which the selection of pig lives has been part and parcel of enhancing human health and welfare. Consequently, the great veterinary expertise on pigs in the country, the accessibility of pigs, and the absence of strong public ethical concerns in relation to the use of pigs in industry and research have greatly contributed to the animal's career as a substitute for the human in the production of biomedical knowledge and its translation to the clinic. In this sense, the history of the pig in Denmark is a central engine in the "genealogy of bioavailability" (Cohen 2005, 85) through which the pig has been shaped as an available resource. I now turn to the practices of modulating this resource, allowing it to enter the clinic and becoming *bio*available.

Spatial Arrangements in the Laboratory

The perinatal pig laboratory in Copenhagen is the only one of its kind in Denmark and is internationally acknowledged for its *Preterm Pig Model* (Sangild et al. 2006). The laboratory is located at the university campus a few kilometers from the university hospital. During fieldwork I followed experiments that investigated the relationship between nutrition and microbes in the gut in the period just after birth. The experiment was repeated multiple times, but here I describe only one typical case.

Some days before the experiment, a pregnant sow from a farm outside Copenhagen has been brought to the stables at the university campus. On the first day of the experiment, the sow is anesthetized and moved from the stables to a room in the laboratory where a group of students prepare her for C-Section by shaving and washing her stomach, doing "all the dirty work" as one graduate student summed it up for me. Then the huge 250-kilo body is carried into the operating theater, where professors, doctoral students, animal technicians and master students, who each have specific tasks to do in relation to the C-section, gather around her. The associate professor who leads the surgery instructs her master's students how to cut through the skin, and as soon as the first piglet appears the small talk stops and everyone follows how one piglet after the other is taken out of the large body. The doctoral students start ventilating the piglets and then move them to a separate room where they are placed in heated, ventilated, humidified and oxygenated individual incubators. The sow stays in the operating theater where a group of students are to practice their stitching skills. Later in the day she is killed. In the other room with the piglets, a group of both senior and junior researchers begin the work of providing each animal with a letter, determining their sex, weighing them, and registering all data in the computer. They also put catheters on the piglets and connect them to individual nutrition machines from which they will be fed during their life in the laboratory. The piglets' anesthetized bodies do not object to the laboratory procedures and the researchers' skilled work on connecting the nutrition machines to the piglets' limp bodies foregrounds the animals as *zoe*, as biological life to be exposed to different forms of microbiota within a settled time frame.

This familiar and well-established understanding of the animal as a tool to understand biological mechanisms is well illustrated by one poster that decorated the corridor of the laboratory before it was moved to a new facility. This poster, which is but one of many that the scientists have presented at scientific conferences, describes nutrition and immunology in piglets during the period before and after birth, and shows pictures of full animal laboratory pigs and close-ups of specific organs. On the poster the pigs are placed in relation to nonsentient and nonspecies-specific physiological phenomena and the scientific literature on this. The pigs are faceless and there are no traces of kinship relations or of the environment that they are part of, such as the lab. Emphasis is placed on the pig as a biological system revealing biological information.

The perinatal pig laboratory is located on the ground floor in the building where the researchers have their offices. This spatial organization underlines the configuration of the piglets as belonging to the scientific domus of the researchers. When the researchers in conversation with me comment on the privileged conditions that the piglets have in the lab, compared with the conditions of prematurely born piglets in the industry, the framework of such conversations is that the piglets have forever left the production system and are envisioned in the sphere of science.

The whole process of preparing the sow for C-section and installing the piglets in their incubator illustrates the first step of making the piglets bioavailable. Biologically the exposure to adverse nutritional conditions and microbiological challenges in being born preterm make the piglets sensitive to NEC and hence good models of the disease and its patient. Conceptually and spatially the enrollment of a mundane production animal into the laboratory – moving her from the space of "dirty work" to the operating theater, separating the piglets from the sow, placing them in individual cages – cleanses the sow and her piglets of their origin as production animals and treats them as biological units, as "ungrievable others" (Franklin 2007,168). All acts in the laboratory enact the integration of the laboratory and the life of the singular piglet in order to establish the animal as a resource to provide new insights into important pediatric diseases such as NEC, and contribute to improving feeding regimes and rearing conditions in the critical neonatal period.

Daily Laboratory Practices

At the same time as the laboratory work treats the piglets as unspecified biological life, daily laboratory practices illuminate a much greater complexity. Often it is the same group of 4–5 graduate students who are on watch in the laboratory both days and nights and attend to the piglets with meticulous care. This involves a continuous attention to their well-being as individuals and a readiness to act if a piglet has, for example, bloody diarrhea or superficial breathing. In attending to them individually and commenting on their singularity, the researchers do not simply interpret the piglets from a distance, but "populate the place with them" (Despret 2008, 123–139; see also Svendsen and Koch 2013, 123). In these daily practices, the piglets are both treated as standardized guts, as *zoe*, that react to microbiota, and as qualified and sentient lives, as *bios*, placed in relationship to the researchers who feed them individually, change their diaper-like cloth, and with great efforts reduce the pain they may experience. Although the character and extent of individualized life-maintaining interventions are a continuous discussion among the researchers, they do consider their individual treatment of piglets central to making them survive until the last day of the experiment. This is important as piglets that need to be euthanized "before time" or happen to die by themselves do not make as reliable data as piglets who have made it to the last day of the experiment. This suggests that not only treating the piglets as standardized *zoe* but also providing them with *bios* is crucial in turning them into valuable resources for human health.

In the scientists' most recent efforts to develop an animal model that provides knowledge of the relationship between gut microbiota and brain development in preterm piglets, the piglets were kept and cared for in the lab for up to 26 days. In one experimental week, and as a kind of an amusement, some of the students involved provided each of the piglets with names of Nobel Prize winners. Piglet A became *Albert* Einstein, R became Conrad *Röntgen*, D became Clinton *Davisson* and name signs with pictures of these famous scientists decorated the individual incubators. The names made it easy for the researchers to distinguish one piglet from the other at the same time as the names also jokingly played with the piglets' identity as raw material in the biography of famous researchers, thereby expressing the dual identities of the piglets as both *zoe* and *bios*.

This dual identity did not only manifest itself in the interaction with the live piglets, but also in the spatial practices during "kill day." On the last day of the experiment, the researchers kill the piglets and prepare the samples in an atmosphere of both highly concentrated work and the excitement characteristic of putting an end to something and paving the way for the next work phase, the bench work of analyzing samples. On one such day I followed Anna, who had done many of the cognitive tests on the piglets in the "brain studies" and spent hours with them during their 26 days in the lab. In the room of the live piglets she anesthetized *Davisson* and wrapped the sleeping creature in a cloth and carried it as a little child in her arms. She said to me "it is still my little pig that I have trained, but it has already become something completely different by now when it is anesthetized." On entering the dissection room she carefully put it on a tray and covered its body with a cloth. One of the experienced researchers told me that the cloth protects the piglet from the bright laboratory light, and apart from that "it also just feels right." The practice of moving the piglet into the dissection room and preparing it for killing at one and the same time treated it as a sentient being ("my little pig" to be wrapped and protected) and literally hid its sentience by turning it into a faceless laboratory object ("something completely different" to be covered by a cloth) upon arrival in the dissection room.

Four people were busy equipping the room of dissection. They collected and put labels on cassettes, prepared the scoring sheets on which they were to note the condition and weight of the different organs of the animals, collected the knives they would need and prepared the nitrogen tank and put it on the table. As soon as they had finished, one of the experienced researchers moved the anesthetized piglet from the tray and killed it with a needle in its heart. The researchers began the concentrated work of cutting it up, stuffing it with ice to keep the tissue fresh, cutting out the different organs, weighing them and placing tissue parts in the small plastic containers which were then dropped into the liquid nitrogen tank. Each piglet was turned into 70–80 samples marked with a number with which it would be known in all future work done on the samples. Gone were the letters and the names that the researchers had used.

Or not quite. At some stage in this intense work process, Anna entered with a new anesthetized piglet in her arms and stopped to chat with the researchers who were cutting up the brain. One of them detached the brain from the skull and placed the empty skull on the table while he concentrated on the important work of cutting out samples of the brain. Looking at the skull intact with eyes and hair, Anna recognized that this was *Röntgen*. She grabbed her mobile phone and showed us a one minute video of *Röntgen* which she recorded a few days earlier. On the video *Röntgen* was eagerly running around and sniffing in its cage.

The incident illustrates that it is not only in interaction with the live piglets that the piglets momentarily come to appear as *bios*. The biographical dimension is also woven into the space of carving out scientific samples (i.e. "*Röntgen* was happily sniffing two days ago and is now an empty skull on the table"). Hence, we may conclude that throughout the whole laboratory process, the process of making bioavailable relies on constituting the piglets as *zoe* yet momentarily allowing them to appear as *bios*. Bioavailability rests upon a slippage between the piglet as *zoe* and as *bios*.

Scientific Presentations of (Pig) Research

The importance of space in allowing piglets to appear as sentient beings and not only biological life was brought home to me when I participated in research seminars with the group of people I had followed in the lab. Tellingly, at these seminars outside the laboratory, the continuous work of turning pigs into instruments for human health did not include the *bios* dimensions of their lives. In the researchers' presentations, the piglets appeared in the language of medical categories such as "gut microbiotal colonization," "mucosal inflammation," "intestinal proteome changes." At these seminars, researchers from the dairy industry as well as clinicians were present and discussions were initiated about research results in the context of clinical care and the development of new milk products for newborn infants. While the researchers often showed photos of the animals in the lab in their presentations, they did not introduce their listeners to the hard work of caring for individual animals. In one such presentation, the researcher stated that "we are going to do another long-term study to reach a greater number of pigs" (and thereby a greater statistical evidence), a description in which the piglets appeared as simply data. During a walk along the harbor close to the conference site in one of the breaks I talked to the same researcher about "another long-term study." In this space he told me about the long work hours during long-term experiments, the husbandry it involved day and night when handling strongly compromised piglets that are all different, and the experience that other research tasks were pushed aside during pig studies. Situated outside the room of presentations, what became important to know about "another long-term study" was not only animals as more data, but also all the laborious work of making the piglets viable within the experimental apparatus.

Likewise, it was on the back seat in the intimate space of the car when going back from a conference that I engaged in conversations with researchers about handling animals in the laboratory. In this private space separate from the auditoriums, stories were told about the procedures as well as the emotions related to caring, killing and dissecting animals.

At the same time as the daily laboratory practices shatter a sharp distinction between the pig as *zoe* and as *bios*, a spatial and conceptual segregation operates between these two identities of the pig as soon as laboratory results leave the lab and move into the semi-public spaces of translational medicine. In conference spaces the piglets as *bios* may be conceptualized as what Wenzel Geissler (Geissler 2013, 13–34) calls an "unknown known," that is, an aspect of reality that is open to experience, but actively not known by way of, for instance, oversight or discursive conventions. Based on many years of fieldwork in an African city and involvement in one large transnational research site, Geissler describes how research staffs' knowledge about research subjects' hunger and lack of health care was central to making the clinical trial function on a daily basis (sharing food with research subjects, convincing them to join and stay in a trial), yet all this knowledge did not constitute "data" and was excluded from scientific reports. In the context of transnational clinical trials, this exclusion contributed to upholding the principal values that underlie research subjects' participation, such as autonomy, and adhering to scientific conventions according to which hunger belongs to a background reality outside the focus of research (ibid., 18–20). Geissler convincingly demonstrates that this form of unknowing should not be understood as a lack of knowledge or as an act of suppressing knowledge. Rather researchers oscillate between knowing and unknowing and thereby productively link bodies, lives and institutions (ibid., 28).

In the case of translating pig studies into human health, the researchers' switching between knowing the piglet as *bios* in the laboratory and unknowing it in the sense of excluding it from public conversation at seminars and conferences was essential to making the piglets bioavailable: knowing the piglets as not only standardized guts, but also sentient beings and treating them as individuals in the laboratory made it possible to create good data without which the research would be impossible. At the same time, in treating the piglets as replaceable biological units when abstracting them into medical categories the scientists created recognizable knowledge within the scientific community and facilitated the piglets' attachment to the clinical treatment of NEC or the development of new milk products for neonates. In this setting, to describe the care practices and present a piglet as "Röntgen" would certainly constitute human/nonhuman relationships "out of place" (cf. Jones 2000). In the short time available in a paper presentation, the scientists did not see information about daily care practices as crucial for a discussion around identifying the optimal time, amount and composition of nutrition to neonatal infants at risk of NEC. When silencing the care practices and the piglets as *bios*, the researchers adhered to ideals of standardization, detachment and reproducibility in the biological and biomedical sciences. They placed the piglets within the scientific field of pediatric gastroenterology and productively linked scientists, institutions and industrial partners in translational medicine. That is, to facilitate the much desired connections between animal-based science and the human clinic, the individual live piglets as well as knowledge about their subjectivity had to be confined inside the laboratory. To explore further the role of these spatial segregations in the infrastructure of translational medicine I will end by making a short excursion to the NICU, the site inhabited by the infants whose health is seen as the ultimate goal of the sacrifice of the piglets.

Spatial Arrangements in the NICU

When a child is born prematurely at the university hospital in Copenhagen, the infant will be rushed to the NICU. The father may decide to stay with the mother who may still be under treatment, or follow the incubator with the infant into the NICU. Here the parents will have a bed next to the incubator and be expected to be present in the clinic during the entire stay of the infant. Soon after the arrival of the infant, the nurses will make a notice providing the infant with a number that identifies it within the institution and states the name of the infant and his/her parents and siblings. Unlike the piglet, the infant not only gains an identity of the institution, but is also placed in kinship. With kinship, the infant becomes an individual (Strathern 1992), just as kin gain new identities of mother, father, sibling, or grandparent with the birth of the child. In the following days, weeks and months, the space around the infant will be decorated with greeting cards, drawings from older siblings, teddy bears and other objects. These objects link the infant to spaces outside the hospital – the homes where parents and older siblings, close relatives or friends have their everyday lives (Svendsen 2015). The biographical aspects of the infant also gain presence through articulations of kinship before any decorations enter the incubator. One day when I followed a nurse attending a very sick child born at week 25 two days earlier, she freed him from all the wires he was entangled in and his mother commented with a smile: "he takes after his father who is such a mess with computers and has wires all over the place. We have got wireless network at home now."

Connections between infants and parents are also established by a body-to-body contact. Usually the nurse will organize that infants, also very small and premature ones, come to lie skin-to-skin on the belly of their mothers and fathers for a short while before entering the incubator again. This is talked about as "getting out" (*at komme ud*) and is usually an event which the parents are excited about and look forward to. "Getting out" also marks that the infant is on its way to graduate to more "open" beds and a life where it can freely be moved around and come close to its parents (cf. Layne 1996). In this way "getting out" enacts what treatment in the NICU is aimed at, namely making it possible for the infant to enter the place of the home and start a life outside the hospital where parents, siblings and friends have their everyday lives. "Getting out" celebrates the infant as one who, unlike the pig, is only temporarily caged in a high-tech environment (the incubator).

The cloakroom of the NICU is decorated with posters, made by parents of admitted children, which with photos and short notes depict the stories of children and the trajectory of their treatments. These posters all start by telling the gestation day and weight of the child, what happened to him/her during his/her time in the hospital – in particular the critical and the joyful events – and how in the end the child in alliance with experts conquered the border zone and entered the home and a normal life with its parents. While the poster in the pig laboratory places the pig in a space of nonsentient nonspecies-specific physiological phenomena, the posters in the NICU place the infants in affective relations that provide a home and a future for them.

This hoped-for take-home baby also informs decisions about life in the NICU. While by far, most infants are successfully treated, a small group of children at the margins of life raise questions about whether or not to extend physical care. In following discussions around such infants, it has been my experience that questions regarding the value of life arrive the moment a child cannot become a take-home baby and cannot be placed within an open-ended temporal framework. When infants were so seriously ill that a graduation towards death within the near future could be predicted, the dominant position among the clinicians was that the most ethically right act towards the child, his/her parent and the public purse would be to discontinue treatment. In listening to conversations among the clinicians around overwhelmingly ill infants, I became aware of the central role of envisioning a human who had an open future and the mental and physical capacities of moving in an open space when it came to deliberations about continuing or withdrawing treatment. In general, infants who could not become a take-home baby and appeared as caged in both space (the incubator) and time (a graduation towards death within a short time) came to resemble the research piglets in their absence of a home and an open-ended temporality (Svendsen 2015). These negotiations about life's worth in the NICU expose central notions of what it means to be a qualified biographical life: To be situated in (kinship) relations and to be able to act freely in space and time (Ferguson 2013). Thereby my ethnography from the NICU also helps to uncover the cultural production of *zoe* in the laboratory, that is, the practices that contribute to uphold a moral division between animal and human essential to establishing piglets as resources for human health. Making piglets bioavailable is thus enabled by *housing* piglets in the state institution within a closed time and by linking viable infants to a *home* outside state institutions and to open-ended futures. The constitution of life as *zoe* or as *bios* is a spatial and temporal achievement.

The simultaneous inclusion of the piglet as *bios* in the laboratory and the exclusion of this dimension outside the laboratory serve to establish the piglets as both morally "like" and "unlike" the humans they represent. Caring for the piglet as a sentient being within the enclosure of the laboratory is pivotal to making it live to the end of the experiment, yet silencing and concealing this dimension outside the laboratory productively make the piglet bioavailable for clinical and industrial players under conditions that are political (promoting human health), economic (generous funding of translational medicine) and moral (claiming the animal as a resource for human health).

Connections and Disconnections between Spaces and Species

In her study of Dolly the sheep, Sarah Franklin notes that Dolly encompasses a human-*as*-animal potential, a human-*via*-animal situation and a human-*versus*-animal history (Franklin 2007, 201). These terms are helpful when discussing the spatial arrangements in the infrastructure of translational medicine in my case study.

The ethnography from my first site, the pig laboratory, illuminates that the human-*as*-animal potential is foundational to the laboratory practice that is based on the idea of a shared biology between piglet and infant due to evolutionary and genetic continuity between humans and pigs.

At my second site, the scientific conferences where research results are discussed with clinicians and the dairy industry, these translational research activities depict and endorse a human-*via*-animal imaginary. The scientific conferences illuminate that it is when transformed into excel sheets, graphs, images and medical categories that the "pig" paves the *via*, the road, between the laboratory and the initiation of clinical studies. Although proximity between laboratory and clinic is one of the central ideas of translational medicine, grunting pigs do not enter the hospital building and undergo C-Section next to women in labor. Neither fleshy pig bodies nor stories about the biographies of piglets in the laboratory move into the semi-public spaces of translational science.

The ethnography from my third site, the NICU, uncovers how the *versus* penetrates all the spatial arrangements of the infrastructure. My somewhat unusual comparison between the spatial arrangements in the pig laboratory and the human NICU elucidates how housing piglets and homing infants constitute the animal as a resource for the human. In the same way as the invisibility around production pigs in Denmark is part and parcel of pigs' significant shaping of the Danish landscape and economy, the public invisibility around sentient piglets in the laboratory is at the core of translating piglets into hopefully significant and visible infant health.

While effective diffusion of knowledge between laboratory and clinic is the basic principle of translational medicine in a research environment devoted to knowledge, my case study highlights that segregation and silencing are as important to making piglets bioavailable. Spatial segregations and "unknown knowns" complete the purification work (Latour 1993) that maintains the physical and moral boundaries that are so essential in connecting the two beings in the battle for infant health and potential. My journey through the infrastructure of animal-based translational medicine illuminates that creating zones of indistinction between animal and human as well as distinct boundaries is integral to paving the piglet's way across spaces and species to the NICU.

Acknowledgments

I am grateful to the researchers and clinicians who allowed me to follow their daily practices and engaged in conversations with me. For stimulating discussions on animal housing, I thank the group who participated in the workshop "Animal Housing: Practical Infrastructures and Infrastructural Practices" in Oslo in 2013. For very helpful comments on an earlier draft of this chapter, I thank Lene Koch, Mie Seest Dam and the two editors.

Note

1 My description of "The study" and "Pigs in Danish Biomedicine" partly overlaps with sections of another article (Svendsen 2015).

References

Agamben, G. (1998) *Homo Sacer: Sovereign Power and Bare Life*. Stanford, CA: Stanford University Press.

Bharadwaj, A. (2008) "Biosociality and Biocrossings. Encounters with Assisted Conception and Embryonic Cells in India," in S. Gibbon and C. Novas (eds), *Biosocialities, Genetics and the Social Sciences: Making Biologies and Identities*. London: Routledge, pp. 98–116.

Buller, H. (2013) "Progress in Human Geography," *Animal Geographies I*: 1–11.

Cohen, L. (2005) "Operability, Bioavailability, and Exception," in A. Ong and S.J. Collier (eds), *Global Assemblages: Technology, Politics, and Ethics as Anthropological Problems*. Oxford: Blackwell.

Candea, M. (2010) "I Fell in Love with Carlos the Meerkat: Engagement and Detachment in Human-Animal Relations," *American Ethnologist*, 37 (2): 241–258.

Davies, G. (2000) "Virtual Animals in Electronic Zoos: The Changing Geographies of Animal Capture and Display," in C. Philo and C. Wilbert (eds), *Animal Spaces, Beastly Places: New Geographies of Human-Animal* Relations. London: Routledge, pp. 243–267.

Despret, V. (2008) "The Becomings of Subjectivity in Animal Worlds," *Subjectivity*, 23: 123–139.

Ferguson, J. (2013) "Declarations of Dependence: Labour, Personhood, and Welfare in Southern Africa," *Journal of the Royal Anthropological Institute*, 19 (2): 223–242.

Ferry, E.E. and Limbert, ME. (2008) *Timely Assets*. Santa Fe: School for Advanced Research Press.

Franklin, S. (2007) *Dolly Mixtures: The Remaking of Genealogy*. Durham, NC: Duke University Press.

Friese, C. (2013) *Cloning Wildlife. Zoos, Captivity, and the Future of Endangered Animals*. New York: New York University Press.

Geissler, P.W. (2013) "Public Secrets in Public Health: Knowing Not to Know while Making Scientific Knowledge," *American Ethnologist*, 40 (1): 13–34.

Groenen, M.A. (2012) "Analyses of the Pig Genomes Provide Insight into Porcine Demography and Evolution," *Nature*, 491: 393–398.

Jones, O. (2000) "(Un)ethical Geographies of Human-Non-Human Relations," in C. Philo and C. Wilbert (eds), *Animal Spaces, Beastly Places: New Geographies of Human-Animal Relations*. London: Routledge, pp. 268–291.

Kirksey, S.E. and Helmreich, S. (2010) "The Emergence of Multispecies Ethnography," *Cultural Anthropology*, 25 (4): 545–576.

Kohn, E. (2007) "How Dogs Dream: Amazonian Natures and the Politics of Transspecies

Engagement," *American Ethnologist*, 34 (1): 3–24.
Kuzmuk, K.N. and Schook, L.B. (2011) "Pigs as a Model for Biomedical Sciences," in M.F. Rothschild and A. Ruvinsky (eds), *The Genetics of the Pig*. Oxford: CABI, pp. 426–444.
Latour, B. (1993) *We Have Never Been Modern*. Cambridge, MA: Harvard University Press.
Layne, L.L. (1996) "'How's the Baby Doing?' Struggling with Narratives of Progress in a Neonatal Intensive Care Unit," *Medical Anthropology Quarterly*, 10: 624–656.
Marsland, R. and Prince, R. (2012) "Introduction: What is Life Worth? Exploring Biomedical Interventions, Survival and the Politics of Life," *Medical Anthropology Quarterly*, 26 (4): 453–469.
Mol, A. (2013) "Mind your Plate! The Ontonorms of Dutch Dieting," *Social Studies of Science*, 43 (3): 379–396.
Netz, R. (2004) *Barbed Wire: An Ecology of Modernity*. Middletown, CT: Wesleyan University Press.
Neu, J. and Walker, W.A. (2011) "Necrotizing Enterocolitis," *New England Journal of Medicine*, 364 (3): 255–264.
Sangild, P.T., Siggers, R.H., Schmidt, M., Elnif, J., Bjornvad, C.R., Thymann, T., Grondahl, M.L., Hansen, A.K., Jensen, S.K., Boye, M., Moelbak, L., Buddington, R.K., Weström, B.R., Holst, J.J. and Burrin, D.G. (2006) "Diet- and Colonization-Dependent Intestinal Dysfunction Predisposes to Necrotizing Enterocolitis in Preterm Pigs," *Gastroenterology*, 130 (6): 1776–1792.
Sangild, P.T., Thymann T., Schmidt M., Stoll, B., Burrin, D.G. and Buddington, R.K. (2013) "The Preterm Pig as a Model in Pediatric Gastroenterology," *Journal of Animal Science*, 91 (10): 4713–4729.
Sharp, L.A. (2013) *The Transplant Imaginary*. Berkeley: University of California Press.
Strathern, M. (1992) *Reproducing the Future*. Manchester, UK: Manchester University Press.
Svendsen, M.N. (2011) "Articulating Potentiality: Notes on the Delineation of the Blank Figure in Human Embryonic Stem Cell Research," *Cultural Anthropology*, 26 (3): 414–437.
Svendsen, M.N. (2015) "Selective Reproduction: Social and Temporal Imaginaries for Negotiating the Value of Life in Human and Animal Neonates," *Medical Anthropology Quarterly*, 29 (2): 178–195.
Svendsen, M.N. and Koch, L. (2013) "Potentializing the Research Piglet in Experimental Neonatal Research," *Current Anthropology*, 54: 118–128.
Wahlberg, A. (2008) "Reproductive Medicine and the Concept of 'Quality'," *Clinical Ethics*, 3: 189–193.
Waldby, C. (2002) "Stem Cells, Tissue Cultures and the Production of Biovalue," *Health: An Interdisciplinary Journal for the Social Study of Health, Illness and Medicine*, 6 (3): 305–323.

13 Closing the Barn Door

Henry Buller

Housing

Houses are places to be, where the micro-conditions for living, resting and interacting are generated and maintained. They are places of protection; protection from predators and from more elemental assaults such as meteorological extremes. They may also be places where prey is trapped and consumed or unpackaged and cooked. Birds construct their nests to protect the young once born and the mother bird during incubation. The nests' location, and sometimes their construction (materials, design), prohibit, or at least reduce, the likelihood of predation. A nest is a (temporary) "home" in the sense that it might encompass a social group, it is a place to which the adults and the young "return" (or "home") for comfort, protection, un-threatening interaction, food and so on. In the UK, there is an additional layer of protection at work. It is an offence, under the 1981 Wildlife and Countryside Act, (for humans) to intentionally destroy or damage the nest of any wild bird while it is in use or being built.

Some migrating animals, birds and fish return cyclically to a specific location – a more spatially extensive "home" place. Yet, while many animals build homes, by no means all occupy them in any permanent sense. In our fiction, Badger's houses may have door scrapers and Toads grow up in Halls, but, as von Uexküll (2010) reminds us, in their "natural" state the animal's house is often mobile or transitory; a temporary site of nurture, safety or defense, an opportunist and transitory halt, where food might be more abundant and some security is to be found in what is often an otherwise itinerant life. Describing the nest-building behavior of chimpanzees, Goodall (1962) observes the construction of new nests each and every night, the fresh foliage of the tree-nests offering greater warmth and comfort. A single great ape, estimate Fruth and Hohmann (1994), may build as many as 19,000 shelters in a lifetime. Burrowing and nesting animals create houses of variable longevity – the naked mole-rat (*Heterocephalus glaber*) may remain located in an elaborate burrow system all its life, like many insect species – while ranging mammals, such the reindeer or the lion make little provision for semi-permanent shelter of even the most temporary kind other than the occasional selection of an appropriate nesting or birthing site.

However temporary, these "houses" are nonetheless "beastly places" where intraspecific communication is enabled, where infants learn from adults and each other (Hansell 2005), where sociability is generated and social evolution enabled (Hansell 1993). Housing becomes a critical reinforcing act of internalizing and externalizing, selfing and othering, defining the line between familiar and the unknown, the wanted and the unwanted, the controlled and the uncontrolled. For human and other animals, the "home" can be the sovereign locus of individual territoriality.

When we humans "house" animals (the foremost act of domestication), we undertake a form of anthropocentric spatial fixity, the creation of an expressly "animal space," as distinct from a "beastly place" (Philo and Wilbert 2000); a place of confinement, of de-naturalization yet one that can also be, to a degree, an act of "humanization" (Nast 2006); a "zoo-ontological production" (Shukin 2009). The mobility and spatial fluidity of the "natural" and wild animal, unharnessed and uncontained, does not always serve us well. Though, increasingly, it is not only domesticated animals that we "house," whether on the farm, in the zoo or in our homes, but wild animals too into defined areas of carefully bordered and protected "naturalness". Whatmore and Thorne (2000) demonstrate that even the wilder spaces along with their occupants have long been enrolled in a relational topology that is a form of both semiotic and material placing. Some animals house themselves, some animals have housing thrust upon them, others (an increasing number) are born housed.

For some "pet" or "companion" animals in Western societies, the outside kennel has been gradually been replaced as the quintessential animal "house" by shared human–nonhuman indoor living spaces, extending to shared beds and eating-places. The human home/house is redefined by animal presence. Jean Cocteau loved his cats because, in his words, "I enjoy my home; and little by little, they become its visible soul" (Cocteau 1951, 568). Even Deleuze, infamous for his "anyone who like cats or dogs is a fool" (Deleuze and Guatarri 1996, 265) readily admits to sharing his home with a number of cats since the "fatal moment," when his son brought home a stray (Deleuze 1987). In a very literal re-territorialization, these are now "indoor" animals with little, if any, hope of rehabilitation.

Across the world, zoo animals live in – or rather, are confined to – artificial and artefactual dwellings (the "Lion House", the "Monkey House") whose design frequently reflects fundamental human framings of both animal and human worlds (Mullan and Marvin 1998). Here, protection and prohibition must co-exist with visibility, with accessibility and with (sometimes almost continual) interference. Although these are presented as the animals' "houses", with nowadays, increasing reference to "natural habitat", they are often, by definition, very far from "home."

Farm animals are "housed." The word now takes on a very different meaning. The word "house" is singularly inappropriate. There is little of the "house" and even less of the "home" in a farrowing stall, a battery cage, a slaughter pen or even an industrial broiler shed. "Housing" becomes an integral and intentional part of the controlling production environment and process; it brings animals, machines, technologies and buildings together in a vocational and techno-scientific materiality that is – in certain circumstances – almost indissociable from that of the animals themselves. In these artificialized spaces, social groups can be utterly re-defined as breed or species groups, often of a single gender, age and size. Family groups are re-constituted. In her film "Blackfish," Cowperthwaite, shows how the so-called families of killer whales in the Sea World parks are not related individuals but assembled collections of different ages from different sources. A tiger in a zoo display may urinate against the concrete walls in a vain attempt to demark the memory of once-guarded and patrolled territories but more often than not, it is in response to some artificially introduced olfactory stimulation (such as chilli, cinnamon or cumin powder).

The worlds of these farmed and to a lesser extent zoo animals, like their bodies, even their genes, are no longer their own. We sculpt animal bodies to fit into their housing and adapt animal behavior so they can live within their "housed" environment. Circularly, the animals' physiological and psychological ability to cope with that housing come to define and determine animal welfare. In this, the paradigms of protection and prevention become hardened under a logic of near-total control: via the protection (and enhancement) of productive body processes through which the very materials and forms of housing units encourage greater productivity and via the prevention of "outside" elements, be they microbial or meteorological that might slow or deviate those processes. Perhaps, the ultimate expression of productive animal housing is in-vitro meat where all the un-necessities of the *bios* are removed.

As the editors of this volume point out in their introduction, the forms and practices by which humans house animals reveal complex concerns. At one level, these are, undeniably, inevitably and intentionally instrumental and asymmetrical spaces, places and structures of control in which human interests take precedence. Yet to see them entirely in this light is to miss something. All of these "housed" animals are still living, mobile, sentient and troublesome "critters" whose interwoven lives, and whose living of those live matters to them and should matter to us. The farm is, as Porcher (2001) maintains, a space of interacting human and nonhuman *umwelts*, where both are – albeit differentially – housed. Current interest amongst human–animal scholars and amongst animal welfare scientists for attention to be paid to a more enlightened and relational accounting of the shared dwelling spaces, materialities and co-presences of the farm, the home, the zoo and even the "wild" (Kirksey 2014), suggest that too instrumentalist and mechanistic a view of the purely physical and environmental parameters (including housing) of animal lives might obscure alternative approaches to the understanding of those "naturalcultural contractual arrangements" (Haraway 2008) of intra- and inter-species flourishing.

In a couple of papers, Bruno Latour (1998; 1995) considers doors, or more precisely a cartoon cat flap built for a cartoon cat into a cartoon door and an artificial, nonhuman hydraulic "door closer." In the former, the series of modifications made to the door to accommodate the shifting priorities of human and nonhuman actors illustrate the creative dance of redefining relational interactions between people, things and animals. In the latter, the automatic door closer shapes, at least a little, bit, the worlds of those who pass through. For Latour, the essence of both these devices or techniques is "the mediation of the relations between people on the one hand and things and animals on the other" (Latour 1995, 1). The technical, the nonhuman and the human are all social actors and it becomes "useless to impose a priori divisions between which skills are human and which are not human, which characters are personified and which remain abstract" (Latour 1998, 305).

> No one has ever seen a technique, and no one has ever seen a human. We only see assemblies, crises, disputes, inventions, compromises, substitutions, translations, and orderings that get more and more complicated and engage more and more elements.
>
> (Latour 1995, 6)

How then do the multitudinous physical structures and techniques of animal housing contribute to such assemblies? That, after all, has been the central question of this entire volume. In the closing chapter, I want to think about how various housing practices and structures both derive from, yet also trouble, the relations we humans might have – or think we have – with sentient, yet domesticated, nonhumans. In the final section of this chapter, I want to imagine different endings. For many millions of domesticated animals, the final "house" is the slaughterhouse where materialities are re-fashioned. What possibilities, if any, do these particular houses offer for anything more than absolute and violent domination?

The Cat Litter Tray

The cat litter tray or litter box has become an essential convenience in the management and practice of contemporary human–animal relations. These plastic trays that are found in countless houses and apartments across the globe are typically filled with clay, silica gel, corn or wood chip, in which pet cats are encouraged to both urinate and defecate. Cat litter trays have a number of different forms, some open, some covered, depending largely upon the sensibilities of either the cat, or the owner, or in some cases both. A quick visit to any local pet store reveals a variety of levels of sophistication in tray design, principally over the methods of emptying the soiled litter once the cat had used it.

> There should be a sufficient number of litter trays, at least one per two cats and preferably one per cat, sited away from feeding and resting areas. The trays should be positioned in a quiet place in the house and cleaned at least once a day. Cats with easy access to the outdoors may not need a litter tray, although older cats, those who do not like to go out in bad weather, and cats that are unwell may require one. Cats can have individual preferences for

> litter characteristics, so it may be necessary to provide a range of litter types and designs of litter trays.
>
> (Rochlitz 2005, 101)

Cat litter trays or boxes are usually made of molded plastic. Manufacturers recommend that the boxes or trays be around one and a half times the length of the cat with sides between 12 and 12 centimeters in height (too high and the cat won't use them, too low and they will scratch the litter out all over the floor). Some are covered; most are open. Some have pressure pads and alarms to notify "owners" when they are in use. Cat litter trays have become a vital, albeit temporary, form of animal housing and in their way have made a highly significant contribution to human–animal relations.

The invention or use of this device, which appeared in the first half of the twentieth century, is almost singularly responsible for the explosive growth of pet cat keeping in central urban locations and, in doing so, has helped to reconfigure human–animal relations on a major scale. Up until the 1940s, most cat owners had gardens or at least access to public open space (Rochlitz, 2005). Cats were let out to roam, like the felines of that wonderful 1960s cartoon series "Top Cat". Unneutered cats, difficult to keep indoors in any case, roamed the streets and gardens of cities. As Katherine Grier writes in her book on the history of American petkeeping "the popularity of cats as pets was compromised by the fact that their owners faced the unpleasant problem of life without cat-litter" (Grier 2006, 77). Although desperate and enterprising owners traditionally used newspapers or sand, it was not until the late 1940s that a commercially available clay-based cat litter called "Kitty Litter" became widely available, at least in the US where it was invented by Edward Lowe in 1947 and originally sold in 5lb bags.

With this invention, and subsequent derivations, people could keep pet cats indoors all the time even in the denser housing projects and high-rise apartments, never needing to let them outside. But what if the cats wanted to go outside? With the increasingly popular and cheap practice of neutering male cats (around 80% of domestic male cats are currently neutered), making them a lot more docile and thereby suited to an entirely indoor life, an entirely new configuration of the species *felis catus* emerges, served by an equally docile human population, now charged with the multiple tasks of supplying them with ready-prepared food, occasional affection and the regular removal of used cat litter. Writing in 1988, Mertens and Schar advise that the indoor cat should have at least two rooms in which to roam, while Crouse et al. (1995) recommend a ready availability of cushions.

Alongside the cat tray, new human and cat social practices (and even personalities; Perrine and Osbourne 1998) develop in which both cats and humans play a key relational role (Mertens 1991; Stammbach and Turner 1999; Fox 2006; Bernstein 2007). Franklin (2007) reminds us that cats do not meow to each other, only to us humans. Derrida's cat, that "irreplaceable living being", shared his bedroom and bathroom. Their conviviality ("*being huddled together*" or "*being with*", Derrida 2008, 10) allowed them both to look at each other.

The cat-litter tray may become the particular site of disruptive animal agency. Not all cats use the tray. A large number of many websites are dedicated to this critical societal issue. Using the tray to defecate and urinate has become so normalized in human expectation that not to do so is taken as a sign of feline unconformity, rebellion or chronic unsociability. Of course, in our unevenly relational world, that usually means, for the cat, a one-way trip to the vet or animal shelter. Cats do not use the tray because of behavioral aversions, medical reasons, because they don't like the litter, because they don't like the location of the box and so on. The box or tray becomes a device for the expression of their subjectivity. As one website advises: "If your cat associates her litter box with unpleasant things, you can work to help her develop new and pleasant associations."

The final point I want to consider relating to cat-litter boxes as physical infrastructure concerns their role in the more material biotic pathways and practices that link humans and nonhumans. Cat feces are a well-known source of enteropathogenes such as toxoplasmosis and, less commonly, toxocariasis, both of which have significant implications for human health (particularly for unborn children) and are acknowledged to be on the rise within domestic households. The presence of cat feces in the home (i.e. in the litter tray) is clearly a source of potential transmission. However, the feces are not generally contagious for the first 48 hours after excretion and to avoid transmission, best practice is to empty the tray as soon as possible. Unfortunately, the social practices that have developed around indoor pet-keeping tend to mitigate against this with home-owners preferring generally to leave feces to dry out before removing and changing the litter. This, many feel, also helps the cat locate the litter tray for further use. The result is a clear increase in the potential for zoonotic infection yet pet-keeping, as a social practice, comes with, for many pet keepers, a sort of systematic denial of the existence of such zoonotic pathways as if cross-species infection was just something that is not done in the modern extended multispecies family

The Farrowing Crate

For this second example, I want to draw upon Dawn Coppin's (2003; 2008) illuminating analysis of the farrowing crate in intensive pig production and the manner in which, in her words, "the sow's agency as a temporally emergent phenomenon [...] arises in coordination with the physical environment and the human management system" (Coppin 2008, 53). Tracing the development of this particular form of confinement, Coppin demonstrates how the adoption of farrowing crates reconfigures relations between farmer, pig, the materiality of the "crate" itself and the wider public in both foreseen and unforeseen ways.

Farrowing pens, commonly barred-steel-sided boxes around 200–250 × 50–90 cm, each one holding a single sow from late pregnancy to the weaning of her piglets, are commonplace within intensive pig production (gestating crates having been banned, at least within the European Union). For producers, they have historically offered a number of advantages over free-range or open housing systems. They greatly facilitate intensive production methods (with corresponding reductions in individual human labor input) and permit all-year-round farrowing within indoor housing units. Critically, they protect the newly-born piglets from being inadvertently crushed by the sow during the farrowing period, thereby allowing viable increases to be achieved in the number of piglets born to each sow through advances in genetic and reproductive technologies. Here there is a trade-off between the health and welfare of the piglets, which will grow to be sold for slaughter, and that of the sow who is likely to be required to produce at least two litters (of 10 to 12 piglets) per annum for two to three years before culling. Confined indoor systems, with regulated environments, feed and medical regimes allow for more productive sows and larger litters, which then necessitate further elements of confinement.

As Coppin (2008) shows, the agency of the sow is constantly frustrated and thwarted. She cannot turn around, she is obliged to rest on hard surfaces with little or no bedding (as this interferes with the manure management system) leading to sores and injuries. She often lies in her own manure and urine, whether on solid or slatted flooring, and has little opportunity to stand. Finally, in the absence of bedding material, she cannot indulge in nest-building behavior despite a generally high motivation to do so during the farrowing period, nor can she satisfactorily interact with her piglets. The sow's frustration plays out in behavioral responses, from aggression and the savaging of piglets to the gnawing of the crate bars, general welfare decline and still-born piglets.

Drawing on Foucault, Coppin's aim is to show, on the one hand, how the disciplining of sows under intensive regimes creates a new form of domesticated, docile and disciplined animal via what is a fundamentally unsymmetrical human–animal relationship and, on the other hand, to demonstrate how the processes of disciplining and being disciplined also includes farmers, obliged to enter into new animal management practices, and other disciplinary agents, from animal researchers, required to find new parameters for assessing the welfare of the sows, to a vocal animal-welfare lobby increasingly mobilized in the advocacy of free farrowing systems. Thus, the very nature of pig husbandry is altered, as Porcher (2010) has so vividly described, as the particular confinement of the farrowing cage engenders agency (and, in Porcher's study, suffering) in both keepers and kept.

Increasingly, in a number of countries, there are calls for the abandonment of farrowing crates and their replacement by what are known as "free farrowing systems" (FAWC 2015). A different configuration of human–animal and sow–piglet agency emerges from these alternative confinement methods. While the value of crated farrowing in terms of productivity and efficiency is widely acknowledged, free farrowing systems do offer the sows the chance for greater freedom of expression and "natural" behavior as well as greater interaction with their litters, though they may result in higher levels of piglet mortality through maternal crushing. To overcome this, sows are encouraged to be "better mothers" by displaying good maternal behavior, something that is difficult in the conventional farrowing crates (Weschler and Weber 2007). The role and agency of the farmers too is changed. While close access to the farrowing sows might be more physically difficult in free systems, additional attentiveness is often required in free or grouped facilities. Both the sows and the farmers are response-able social agents.

The Final House: The House of Slaughter

In his book *Meat* from 1991, the anthropologist Nick Fiddes wrote that, for many human societies, the killing of animals and the eating of meat symbolizes a (masculine) desire for domination and control over the natural world. To kill and to eat is a form of mastery. Arguably, at least in Western societies, it is a constant signifier of human superiority over "the beast" – a Cartesian de-signification of the animal.

For Fiddes, modernity's domination and control over "Nature" achieves almost its apogee in the violence of contemporary animal killing. Fiddes (1991) parallels the dramatic growth in animal farming and meat consumption with the Industrial Revolution and its aftermath. Perhaps one of the most evocative expressions of this early modernist mastery was contained in those dramatic sections of Upton Sinclair's famous book entitled *The Jungle*, published in 1906, that dealt with the slaughterhouses of Chicago. Later, Derrida (1990) refers to the technological violence towards animals – and the non-criminal putting them to death – as being "vital to our modernity" (Derrida 1990, 953). Modernity, its economies, its technologies and sciences of confinement and constraint, have made the killing of farm animals so frequent, so intense and so integral to modern economies that is has become infinitely banal (Shukin 2009; Pachirat 2011).

The slaughterhouse, with its inverse-Fordist dis-assembly lines, is a temple to progressive modernity. Paula Young Lee (2008, 2) refers to the slaughterhouse as "one of the exemplary institutions of the nineteenth century" and certainly the very name has that ring to it – like "wash-house" or "poor-house". Even today, slaughterhouses remain linear and highly serialized, with compartments and task divisions that strongly resemble those early twentieth-century manufacturing assembly lines.

Slaughterhouses though, are above all places of violence, though they are ethically paradoxical and contradictory places. As Mick Smith (2002) has pointed out, in societies where the intentional killing of animals is generally abhorred, we kill hundreds of thousands each year in these industrialized killing facilities. We deal with that paradox in three ways: first, physically, by removing slaughter from our view in discreet units; second, juridically, by defining the parameters of killing increasingly tightly under the intrinsically ironic banner of "humane-ness" and "welfare" and; third, through a process of the normalization of slaughter practices (which incorporates a rejection of less "modern" practices such as Shekita and Halal slaughter).

What is the role of violence in animal killing and of rendering their bodies eatable? What sort of violence is this? Does the violence reside in the act of killing (is killing inherently violent)? Does it lie in the reasons behind the act (killing to eat or killing to end suffering)? Or rather is it found in the mechanics of the act (with a pill or with a knife) or even in the aesthetics of the act (we might find the physical and material excesses of blunt force trauma unacceptable and yet be more at ease with suffocation – say of a fish)? When does the violence start – perhaps at the very definition of certain animals as "kill-able"? When does it stop (with death or beyond)? Is there ever a less violent way to kill and to eat?

"There is no way to eat and not to kill" writes Haraway. But there are many ways of killing – both direct and indirect – and, we might argue, not all are equally violent. The trouble, for many, is that there is always violence in the process by which we kill the animals we eat. Indeed that violence has become an absolutely critical and systematic element in achieving the transition from the vibrant materiality of the living body to the somehow very different silent materiality of the edible foodstuff. And that violence is always imprinted on that subsequent material form.

The reasons for – the apparent need for this violence are clear. The most obvious is for reasons of health – the health of the consuming body, rather than the consumed body. The dead animal body decays fast: the faster the killing and the proximity of killing to consumption (or preparation), the lower the risk of contamination and infection. So violence is necessary for speedy dispatch. While technology allows us to lengthen the distance (both spatial and temporal) between killing and consumption, the time-scale of processing, from slaughter to processing are, in fact, becoming shorter and shorter. The second obvious reason for the killing of livestock to be inherently violent is the control of supply. Human societies farm and kill animals according to a rhythm of supply and demand. Within the field-to-fork production cycle, slaughter time (like life-time) becomes a significant variable and must be minimized.

A third reason why killing needs to be violent is because it must kill. It cannot wound or mildly discomfort. To kill effectively, the heart must be stopped and brain activity cease. This cannot be achieved through gentle persuasion. The problem for livestock production is that this must be achieved while maintaining not only the integrity of the animal body, both for reasons of value, but also the relative purity of the animal body. Stock animals cannot be "put to sleep" in that gentle metaphoric way we intentionally end the lives of our pet cats and dogs. The residue of the killing chemical would render the animal's body unfit for consumption.

The mode of killing (in other words, the speed, the violence) critically affects the quality of the meat product. An animal that becomes fearful or panicked will release hormones into its body pre-mortem, which taint the meat. An animal that suffers and struggles against death will create damaged muscle tissue and internal bleeding, visible to the consumer. Violence makes economic good sense for the quality of the final product. There are no gentle ways to kill for consumption.

There is also an ethics here, the faster we kill, the shorter the experience of being killed. For philosophers like Peter Singer, this is critical. He has famously argued that logically, were we able to raise and kill food animals entirely without suffering, and ensure that the dead animal is always "replaced" by a live one, there might be no intrinsic moral issue with animal killing for human consumption (Singer 1975). Do we then correlate "violence" with suffering? In their book *The Ethics of What We Eat*, Singer and Mason (2008) approvingly quote the TV chef, Hugh Fearnley-Whittingstall who, in his own book *The River Cottage Meat Book* (2004) writes: "truly wild animals, dispatched efficiently by a good shot, provide us with meat that is perhaps the least ethically problematic of all".

The irony is that with growing societal ethical concern over animal welfare, and the desire to kill well (i.e. without suffering), we have become, in some ways, more "violent" in the manner of animal killing as the focus has been on speed and instantaneity of death. (Here I take a rather traditional notion of violence – which I accept is perhaps a little limited in scope: violence as "physical force"). The force of the percussive bolt, the strength of the electric current and so on… Moreover, this violence is increasingly technologized, undertaken in wholly artificial environments that resemble nothing of Fearnley-Whittingstall's romantic vision of outdoor slaughter, in a field with a rifle.

So, acknowledging, what Fitzgerald (2010) identifies as a growing tension in "post-domestic cultures" between on the one hand, a growing concern for animal welfare and the quality of animal lives and, on the other hand, ever increasing demands for meat, we might wonder (as we change some of the conditions and assumptions of modernity), what might be the nature of the "post-modern" abattoir and the place of violence within it. Might it be made of glass as Pollan (2002) suggests?

> But maybe all we need to do to redeem industrial animal agriculture in this country is to pass a law requiring that the steel and concrete walls of the CAFO's and slaughterhouses be replaced with … glass. If there's any new "right" we need to establish, maybe it's this one: the right to look. […] Were the walls of our meat industry to become transparent, literally or even figuratively, we would not long continue to do it this way. Tail-docking and sow crates and beak-clipping would disappear overnight, and the days of slaughtering 400 head of cattle an hour would come to an end. For who could stand the sight? Yes, meat would get more expensive. We'd probably eat less of it, too, but maybe when we did eat animals, we'd eat them with the consciousness, ceremony and respect they deserve.
>
> (Pollan 2002)

Might a post-modern abattoir be one without animals and without killing or perhaps a place where animals "die naturally" (should we move from carnivorousness to carrion-eaters?). Perhaps, it will be a site where new forms of life are converted to food, be they insects or forms of "mold" or even cellular and microbial "life" from which "meat" is only "grown"…

Abattoirs, farrowing crates, cat litter trays … all the various forms of animal housing, animal confinement and human–animal place-sharing described in this book have generated, and are generated, from a range of shared practices and shared technologies. These in turn come to define the meaning, the agency and the subjectivity of both the keeper and the kept. Houses are, in varying degrees, dialectic spaces, enshrining both confinement and freedom, self and other. They are places of assembly but above all, places of social, human–nonhuman assemblage.

References

Bernstein, P. (2007) "The Human Cat Relationship," *Animal Welfare*, 3: 47–89.

Cocteau, J. (1951) "Du Serieux," *La revue des deux mondes*, Issues 5–8: 568.

Coppin, D. (2003) "Foucauldian Hog Futures: The Birth of the Mega-Hog Farm," *Sociological Quarterly*, 44 (4): 597–616.

Coppin, D. (2008) "Crate and Mangle: Questions of Agency in Confinement," in A. Pickering and K. Guzik (eds) *The Mangle in Practice*. Durham, NC: Duke University Press, pp. 46–66.

Crouse, S.J., Atwill, E.R., Lagana, M., and Houpt, K.A. (1995) "Soft Surfaces: A Factor in Feline Psychological Wellbeing," *Contemporary Topics in Laboratory Animal Science*, 34, 94–97.

Deleuze, G. (1987) *L'Abécédaire de Gilles Deleuze*. Film produced by Pierre-André Boutang in 1988–1989. Available on DVD.

Deleuze, G. and Guattari, F. (1996) *A Thousand Plateaux*. London: Continuum Press.

Derrida, J. (1990) "Force of Law: The 'Mystical Foundation of Authority'," *Cardozo Law Review*, 919: 920–1046.

Derrida, J. (2008) *The Animal That Therefore I Am*. M.-L. Mallet (ed), D. Wills (trans). New York: Fordham University Press.

FAWC (2015) *Opinion on the Use of Free Farrowing Stalls*. London: FAWC.

Fearnley-Whittingstall, H. (2004) *The River Cottage Meat Book*. London: Hodder and Stoughton.

Fiddes, N. (1991) *Meat: A Natural Symbol*. London: Routledge.

Fitzgerald, A. (2010) "A Social History of the Slaughterhouse: From Inception to Contemporary Implications," in *Research in Human Ecology*, 17 (1): 58–69.

Fox, R. (2006) "Animal Behaviours, Post-human Lives: Everyday Negotiations of the Human-Animal Divide in Pet Keeping," *Social and Cultural Geography*, 7 (4): 525–537.

Franklin, A. (2007) "Be[a]ware of the Dog: A Post-humanist Approach to Housing," *Housing, Theory and Society*, 23 (3): 137–156.

Fruth, B. and Hohmann, G. (1994) "Ecological and Behavioural Aspects of Nest Building in Wild Bonobos," *Ethology*, 94: 113–126.

Goodall, J. (1962) "Nest Building Behaviour in the Free Ranging Chimpanzee," *Annals of the New York Academy of Science*, 102 (2): 455–467.

Grier, K. (2006) *Pets in America: A History*. Chapel Hill: University of North Carolina Press.

Hansell, M.H. (1993) "The Ecological Impact of Animal Nests and Burrows," *Functional Ecology*, 7 (1): 5–12.

Hansell, M.H. (2005) *Animal Architecture*. Oxford: Oxford University Press.

Haraway, D. (2008) *When Species Meet*. Minneapolis: Minnesota University Press.

Kirksey, E. (2014) *The Multispecies Salon*. Durham, NC: Duke University Press.

Latour, B. (1995) "A Door Must Be Either Open or Shut: A Little Philosophy of Techniques," in A. Feenberg and A. Hannaway (eds) *Technology, and the Politics of Knowledge*. Bloomington: Indiana University Press, pp. 52–62.

Latour, B. (1998) "Mixing Humans and Nonhumans Together: The Sociology of a Door Closer," *Social Problems*, 35 (3): 298–310.

Mertens, C. (1991) "Human-Cat Interactions in the Home Setting," *Anthrozoos*, 4 (4): 214–231.

Mertens, C. and Schar, R. (1988) "Practical Aspects of Research on Cats," in D.C. Turner and P. Bateson (eds), *The Domestic Cat: The Biology of Its Behaviour*. Cambridge: Cambridge University Press, pp. 179–190.

Mullan, B. and Marvin, G. (1998) *Zoo Culture*. Champaign: University of Illinois Press.

Nast, H. (2006) "Loving … Whatever: Alienation, Neoliberalism and Pet Love in the Twenty-First Century," *ACME*, 5 (2): 300–327.

Pachirat, T. (2011) *Every Twelve Seconds: Industrialised Slaughter and the Politics of Sight*. New Haven, CT: Yale University Press.

Perrine, R.M. and Osbourne, H.L. (1998) "Personality Characteristics of Dog and Cat Persons," *Anthrozoos*, 11 (1): 33–40.

Philo, C. and Wilbert, C. (eds) (2000) *Animal Spaces, Beastly Places*. London: Routledge.

Pollan, M. (2002) "An Animal's Place," in *The New York Times Magazine*. November 10th, 2002. Available online at: http://michaelpollan.com/articles-archive/an-animals-place/

Porcher, J. (2001) *Éleveurs et animaux, réinventer le lien*. Presses universitaires de France.

Porcher, J. (2010) "Contagion de la souffrance entre travailleurs et animaux en production porcine industrielle. Problematiques et Debats," in *Courrier de l'Environnement de l'INRA*, 58: 5–19.

Rochlitz, I. (2005) "A Review of the Housing Requirements of Domestic Cats Kept in the Home," *Applied Animal Behaviour Science*, 93: 97–109.

Shukin, N. (2009) *Animal Capital: Rendering Life in Biopolitical Times*. Minneapolis: University of Minnesota Press.

Singer, P. (1975) *Animal Liberation*. New York: Harper Collins.

Singer, P. and Mason, J. (2008) *The Ethics of What We Eat*. Emmaus, PA: Rodale Publishers.

Smith, M. (2002) "The 'Ethical' Space of the Abattoir," in *Human Ecology Review*, 9 (2): 49–58.

Stammbach, K. and Turner, D. (1999) "Understanding the Human–Cat Relationship: Human Social Support or Attachment," *Anthrozoös*, 12 (3): 162–168.

von Uexküll, J. (2010) *A Foray into the Worlds of Humans and Animals*. Minneapolis: Minnesota University Press.

Wechsler, B. and Weber, R. (2007) "Loose Farrowing Systems: Challenges and Solutions," *Animal Welfare,* 16 (3): 295–307.

Whatmore, S. and Thorne, L. (2000) "Elephants on the Move: Spatial Formations of Wildlife Exchange," in *Environment and Planning D: Society and Space,* 18(2): 185–203.

Young Lee, P. (2008) *Meat, Modernity and the Rise of the Slaughterhouse.* Durham, NH: University of New Hampshire Press.

Index

Page numbers in *italics* indicate illustrations, n denotes an endnote.

For Product Safety Concerns and Information please contact our EU representative GPSR@taylorandfrancis.com
Taylor & Francis Verlag GmbH, Kaufingerstraße 24, 80331 München, Germany

www.ingramcontent.com/pod-product-compliance
Ingram Content Group UK Ltd.
Pitfield, Milton Keynes, MK11 3LW, UK
UKHW022130190926
475702UK00013B/1732

9781138547155